T0135441

Michał Pasternak

Simulation of the Diesel Engine Combustion Process Using the Stochastic Reactor Model

Logos Verlag Berlin GmbH

λογος

Michał Pasternak

Simulation of the Diesel Engine Combustion Process
Using the Stochastic Reactor Model

Reviewers:
Prof. Dr.-Ing. Fabian Mauss
Prof. Dr.-Ing. Roland Baar

Bibliographic information published by the Deutsche Nationalbibliothek

The Deutsche Nationalbibliothek lists this publication in the Deutsche
Nationalbibliografie; detailed bibliographic data are available
on the Internet at http://dnb.d-nb.de

Zugl.: Cottbus-Senftenberg, BTU, Diss., 2015

ISBN 978-3-8325-4310-5

Logos Verlag Berlin GmbH
Comeniushof, Gubener Str. 47,
D-10243 Berlin
Germany

Tel.: +49 (0)30 / 42 85 10 90
Fax: +49 (0)30 / 42 85 10 92
http://www.logos-verlag.de

Abstract

A probability density function (PDF)-based modelling approach is presented for the calculation of combustion, emissions formation and fuel effects in direct injection Diesel engines. The modelling incorporates a zero-dimensional (0D) direct injection stochastic reactor model (DI-SRM) of engine in-cylinder processes and complex reaction mechanisms for the oxidation of hydrocarbon fuels and emissions formation. An emphasis is put on the modelling of mixing time to improve the representation of turbulence-chemistry interactions in the DI-SRM. The mixing time model describes the intensity of mixing in the gas-phase for scalars such as enthalpy and species mass fraction. On a crank angle basis, it governs the composition of the gas mixture that is described by PDF distributions for the scalars. In the 0D computational domain, the PDFs represent the three-dimensional effects of in-cylinder flows, fuel injection and spray formation on the in-cylinder mixture inhomogeneity in actual engines. The modelled mixing time is interpreted as volume-averaged representative mixing time for 0D PDF-based simulations of Diesel engines.

The developed mixing time model, the detailed reaction kinetics and the DI-SRM are accommodated into the engine cycle simulation process. The history of mixing time is derived through an extended heat release analysis that has been fully automated using a genetic algorithm. The predictive nature of simulations is achieved through the parametrisation of the mixing time model with known engine operating parameters such as speed, load and fuel injection strategy. An exact treatment of the non-linearity of reaction kinetics, which governs the local combustion rates and pollutants formation, enables the extrapolation of engine performance parameters beyond the operating points used for model calibration. In turn, crank angle dependency of the mixing time improves the modelling of local inhomogeneities of species mass fraction and temperature of the in-cylinder mixture. In combination with detailed chemistry consideration, it enables an accurate prediction of the rate of heat release, in-cylinder pressure and exhaust emissions, such as nitrogen oxides, unburned hydrocarbons and soot, from differently composed fuels. Exemplary calculations have been carried out for several engine operating points and fuels such as n-heptane, iso-octane, toluene, n-decane and their blends.

Overall, good agreement between the computed and the experimental data for variety of engine performance parameters and exhaust emissions proves the strength of the simulation method developed. The method is particularly tailored for computationally efficient applications that focus on the details of reaction kinetics and the locality of combustion and emission formation in Diesel engines.

Contents

Nomenclature

Abbreviations

ATDC	after top dead centre
CA	crank angle
CA_{10}	crank angle of 10% of the cumulative heat release
CA_{50}	crank angle of 50% of the cumulative heat release
CA_{90}	crank angle of 90% of the cumulative heat release
CAD	crank angle degree
CD	coalescence-dispersion (mixing model)
CDF	cumulative distribution function
CFD	computational fluid dynamics
CHR	cumulative heat release
CN	cetane number
CO	carbon monoxide
CO_2	carbon dioxide
CPU	central processing unit
DI	direct injection
DI-SRM	direct injection stochastic reactor model
EGR	exhaust gas recirculation
EMST	Euclidean minimum spanning tree (mixing model)
EOC	end of combustion
EOE	end of exponential decay of mixing
EOI	end of injection
EOV	end of vaporisation
EVO	exhaust valve opening
GA	genetic algorithm
HC	hydrocarbon (unburned)
HCCI	homogeneous charge compression ignition
IC	internal combustion (engines)
IEM	interaction by exchange with mean (mixing model)
IMEP	indicated mean effective pressure
IVC	inlet valve closure
LHV	lower heating value
MC	Monte Carlo

MDF	mass density function
NO	nitrogen monoxide
NO_2	nitrogen dioxide
NO_x	nitrogen oxides
NTC	negative temperature coefficient
OP	operating point
PAH	polycyclic aromatic hydrocarbon
PaSPFR	partially stirred plug flow reactor
PDF	probability density function
PM	particulate matter
PRF	primary reference fuel
PSDF	particle size distribution function
RoHR	rate of heat release
SOC	start of combustion
SOI	start of injection
SOV	start of vaporisation
SRM	stochastic reactor model

Greek Symbols

α	summation index	–
χ	scalar dissipation rate	s^{-1}
$\dot{\omega}_i$	mass rate of formation of species i per unit volume	$kg\ s^{-1}$
ϵ	turbulent kinetic energy dissipation rate	$m^2\ s^{-3}$
$\mathcal{M}_i$	chemical symbol of species i	–
ω_i	molar net rate of formation of species i	$mol\ m^{-3}\ s^{-1}$
ϕ	generalised flow property or equivalence ratio	–
ψ	sample space variable corresponding to Y_i, M_r, T	–, m^{-3}, K
ρ	density	$kg\ m^{-3}$
τ_ϕ	time scale of the scalar fluctuations	s
τ_t	time scale of the velocity fluctuations	s
τ_{ch}	chemical time scale	s
φ	crank angle degree	deg

Roman Symbols

C	progress variable	–
C_ϕ	mixing time constant	–
C_h	stochastic heat transfer constant	–

c_p	constant pressure specific heat	J kg^{-1} K^{-1}
Da	Damköhler number	–
deg	degree on crank angle basis	deg
E_a	activation energy	J
Exp	experimental quantity	–
h_i	specific enthalpy of species i	J kg^{-1}
$h_{f,i}^0$	specific enthalpy of formation of species i	J kg^{-1}
i	species index	–
J_α^i	molecular flux of species i	mol m^{-2} s^{-1}
k	turbulent kinetic energy	m^2 s^{-2}
l_I	integral length scale	m
m	mass	kg
M_r	soot moment of order r	m^{-3}
N	number of samples	–
N_C	number of cycles	–
N_i	number density of soot particles of size class i	m^{-3}
N_P	number of particles	–
N_R	number of reactions	–
N_S	number of species	–
p, p_{max}	pressure, maximum in-cylinder pressure	bar
Q_i	general source/sink operator for species i	s^{-1}
r_j	reaction rate of the jth reaction	mol m^{-3} s^{-1}
R_u	universal gas constant	J mol^{-1} K^{-1}
Re	Reynolds number	–
Sim	simulated quantity	–
T	mean gas temperature	K
t	time	s
T_w	cylinder wall temperature	K
u	velocity	m s^{-1}
u'	velocity fluctuation	m s^{-1}
V	volume	m^3
W	mean molar mass	kg mol^{-1}
W_i	molar mass of species i	kg mol^{-1}
x_α	spatial position	m
X_i	mole fraction of species i	–
Y_i	mass fraction of species i	–
Z	mixture fraction	–

Chapter 1

Introduction

1.1 Background

Meeting the ever stringent emission standards and improving fuel efficiency of internal combustion (IC) engines requires continuous development of the design methods. These methods are based on the experimental work and benefit from numerical modelling and simulations. Particularly useful are numerical models for the prediction of the rate of heat release and pollutants formation that have fundamental impact on engine efficiency and exhaust emission levels. The use of numerical simulations continues to increase at all phases of engine development. During an early phase, simulations of engine cycle performance are frequently decisive while developing the initial engine concepts. The demand for low computational cost favours zero-dimensional (0D) models of engine in-cylinder processes. Application of these is most effective if both the rate of heat release and engine exhaust emissions are accurately predicted. This is a challenging task that is amplified especially for Diesel engines. Mixture preparation, combustion and subsequently emission formation are locally governed by the instantaneous interactions between the chemical reactions and turbulent flow. These interactions determine mixing between reactants that on a molecular level is the prerequisite to initiate combustion. Mixing has very local character and depends on the geometry effects that cannot be well represented by 0D models. As a result, the modelling of local inhomogeneity of temperature and equivalence ratio, which result from turbulence-chemistry interactions, is limited and inhibits an accurate prediction of the local rates of heat release and pollutants formation. The present work aims at improving the representation of the effects of turbulence-chemistry interactions within 0D modelling framework. The focus is on further development of a probability density function (PDF)-based stochastic reactor model (SRM) for the simulation of combustion, emission formation and fuel effects in Diesel[1] engines. In particular, a variant of the SRM for the simulation of direct injection (DI) engines (DI-SRM)[2] is investigated.

1.2 Simulations in the Development of Diesel Engines

1.2.1 Modelling of Combustion Processes

Depending on the modelling concept and application targets, different types of numerical models are distinguished for the simulation of in-cylinder processes in combustion engines. A detailed review of these models is beyond the scope of this work and can be found elsewhere [96, 133, 146]. This section is confined to a basic review that aims to provide a reference for the SRM-based modelling approach that is discussed in this work.

The dimensionality of the engine in-cylinder processes is a most common criterion of classifying engine numerical models. Here, one distinguishes between three-dimensional (3D) computational fluid dynamics (CFD), quasi-dimensional (QD) and 0D models as proposed in [21]. The 3D CFD models [84, 120, 144, 164] present the most comprehensive description of engine in-cylinder processes. They account for the turbulence effects and allow a detailed study of mixture formation and combustion. However, these benefits are occupied by the complexity of setting up the model and still rather high cost of computations. Particularly expensive can be these applications of 3D CFD models that require complex chemical reaction mechanisms for the modelling of pollutants formation. The QD models, as for example [57, 64, 65, 70, 139, 147, 148], fall into between the 0D and 3D CFD models. For the calculation of the rate of heat release, they account for some physical and chemical processes inside the cylinder, but in a simpler fashion than the 3D CFD models and by omitting the information about the flow structure. Usually, the QD models are formulated as multi-zone. Mostly, for

---

[1]In this work, the term Diesel refers exclusively to Diesel engines with direct fuel injection.

[2]The term SRM is considered as an overall modelling concept, whereas the term DI-SRM emphasis the capability of the model to simulate direct fuel injection.

the simulation of exhaust emissions they use global reaction mechanisms or simple empirical correlations that is their major shortcoming. The 0D models [10, 20, 26, 27, 38, 46, 61, 81, 119, 131] are the simplest in terms of the formulation and hence, most efficient in computations. Usually, they are formulated as single-zone, but may also occur in multi-zone variants [5, 99].

Dimensionless treatment of the in-cylinder processes by the 0D models is their main drawback. It prevents an accurate representation of the locality and the 3D character of turbulence-chemistry interactions that govern the combustion process and emission formation in actual Diesel engines. The difficulty of representing the 3D Diesel combustion process in 0D computational domain can be avoided in the SRM [17, 77, 90, 101, 149, 150, 157] of engine in-cylinder processes. The SRM is formulated as 0D, but the employed PDF methodology [59, 126] enables representing the effects of turbulence-chemistry interactions [39, 47, 91]. This is achieved through the modelling of local inhomogeneity of temperature and species mass fraction in the fuel spray and mixture. Hence, in terms of spatial effects, the SRM falls into between the 3D CFD and the QD models, whereas the computational cost places it between the 0D and the QD models.

1.2.2 PDF-Based Engine Simulations

PDF modelling refers to a probabilistic [126, 127] method of calculating turbulent flows with reactions. In this approach, the properties of the flow, such as velocity, species mass fraction or temperature, are obtained from the PDF distributions. The PDF is a dependent variable that contains a complete statistical description of the state of the fluid at each point and time in the flow field – one-point and one-time PDF.

PDF Method and SRM for Diesel Engines

The two characteristic features of the PDF method are (1) an accurate treatment of the chemical reactions and (2) the modelling of molecular transport [59, 69, 126] to represent the effects of turbulent mixing. These features make the method suitable for the simulation of Diesel combustion.

- Application of the detailed reaction mechanisms for the description of fuel oxidation and pollutant formation allows a more thorough understanding of the formation of engine exhaust emissions such as nitrogen oxides (NO_x), unburned hydrocarbons (HC), carbon monoxide (CO), carbon dioxide (CO_2) and particulate matter (PM). Important intermediate or precursory species and reaction pathways can be tracked. Fuel's resistance to self-ignition can be investigated that is one of the key parameters that governs the combustion process and pollutants formation in Diesel engines.

- By the modelling of mixing process, local inhomogeneity is introduced into the gas-phase for species concentration and temperature. This, in some extent, mimics the turbulence effects and models the heterogeneous Diesel combustion. As a result, the spatial effects of the combustion and pollutants formation processes are mimicked, which is crucial for an accurate prediction of exhaust emissions.

The PDF method is a foundation of the SRM for the analysis of turbulent reactive flows [76], combustion in IC engines [77, 90] or furnaces [9] and a variant of the SRM for the simulation of IC engines with direct fuel injection (DI-SRM) [150, 157]. The SRM/DI-SRM considers gas inside the cylinder as an ensemble of notional particles[1]. The particles represent points in the gas-phase for species mass fraction and enthalpy from which the temperature is calculated. These scalar are treated as random variables and are described by PDF distributions. The properties of particles evolve following the PDF transport equation that for direct injection Diesel engines accounts for the effects due to piston movement, chemical reactions, convective heat loss, fuel injection and turbulence. In more complex versions of the SRM, the flow into and out of the cylinder can be included [87, 150]. Overall, the SRM/DI-SRM can be seen as a simple representative 0D PDF engine model.

SRM in Engine Applications

Previous applications of the SRM were primarily concerned with simulations of various concepts of premixed or partially premixed combustion modes. These studies have demonstrated the importance of detailed chemistry information for the combustion process. In [90], the SRM was applied to study the influence of mixture inhomogeneity caused by the thermal boundary layer adjacent to the cylinder walls on the combustion process in homogeneous charge compression ignition (HCCI) engines. In [101], the SRM was integrated with 1D code GT-Power and applied to the simulation of a HCCI turbocharged engine. Later, in [17], such an integrated tool was used to simulate a dual-fuelled

---

[1]The terms virtual or stochastic particles are also used.

HCCI engine. In [4], the SRM was applied to investigate the formation of NO_x in a HCCI engine operated at different exhaust gas recirculation (EGR) rates, fuel-air ratios and inlet temperatures. In [150], using the SRM, an influence of the relative air-fuel ratio and early fuel injection timing on combustion and emissions formation in an iso-octane fuelled HCCI engine was studied.

In [102], the features of the SRM were further extended by introducing a fuel spray model featuring the injection process. The model was applied to simulate a gasoline fuelled HCCI engine. A similar model has been used in [149] to simulate the effects of two-stage direct fuel injection and fuel/wall interactions in a Diesel fuelled HCCI engine. In [18], the SRM was applied to simulate combustion of a natural gas fuelled HCCI engine.

So far there have been relatively few applications of the SRM for the simulation of combustion processes in purely Diesel engines. In [135] the performance of different mixing models was tested under Diesel conditions. The results obtained revealed the capability of the SRM to reproduce the in-cylinder pressure, soot volume fraction and ignition onset. In [140], the SRM was used to examine the local in-cylinder temperatures and equivalence ratios and to highlight the main sources of excessive exhaust emissions in Diesel engines. In [157], a variant of the SRM for the simulation of direct injection engines was coupled with 1D gas dynamic tool to enable the simulation of a complete Diesel engine cycle. The coupled model was further examined in [111] with respect to the prediction of engine exhaust emissions and global performance parameters such as brake mean effective pressure and fuel consumption.

Modelling of Turbulence Effects

Turbulence effects are included in the SRM through the modelling of mixing between particles. Loosely, the mixing process is understood as a change in time of particle's scalar properties, such as species mass fraction and enthalpy, with the intent to make an ensemble of particles a more homogeneous system as time goes by. The change of the scalar properties is obtained by certain unit operations called *mixing events* that are fully defined by the applied mixing model (see, e.g., Section 3.5.5). The frequency of mixing operations is controlled by the mixing time scale that in this work is referred to as the *mixing time* with a unit of second. With respect to actual engines, the mixing time can be understood as an inverse of the frequency at which air, fuel and rest gases mix with each other. In the SRM, the shorter the mixing time the higher the intensity of the mixing operations on particles and vice versa.

Usually, the mixing process is realised by using one of several existing stochastic or deterministic mixing models such as the interaction by exchange with the mean (IEM) [35], the coalescence-dispersion (CD) [28] or its modified version (MCD) [36, 67], the Binomial-Langevin [160] and Euclidean minimum spanning tree (EMST) [151]. The characteristic feature of these models is the necessity to know the mixing time in advance to simulations. This is a one of the challenges in 0D PDF calculations of combustion engines. In 0D models the 3D geometry effects cannot be well respected that is crucial for an accurate determination of the mixing time.

In the existing applications of the SRM/DI-SRM, the mixing time was mainly modelled by assuming mixing intensity to be homogeneous in space and time during the closed part of engine cycle that is, a single value mixing time was used. This approach was successfully applied to HCCI engines [17, 90, 102, 150].

With respect to the simulation of direct injection Diesel engines, the use of constant value mixing time is questionable. The in-cylinder processes, such as flow pattern, fuel injection, mixture formation and pollutant formation, are characterised by different time and length scales [61, 146]. The scales change during the cycle that corresponds to the changes of mixing intensity and hence, to the changes of mixing time. The temporal and spatial changes of mixing intensity can be deduced from measurements of instantaneous velocity fields in a transparent engine (see, e.g., [31]). Those measurements indicate strong variation of velocity vectors that in some extent suggests an overall intensity of the flow process in the combustion chamber. Further confirmation of mixing time changes during the cycle is provided by 3D CFD computations as for example these presented in Fig. 1.1.

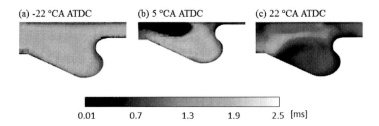

Figure 1.1: Exemplary 3D CFD computations of mixing time at different instants during the cycle from a Diesel engine with direct fuel injection

In Fig. 1.1, the mixing time was calculated as the ratio between the turbulent kinetic energy and its dissipation (see also Section 5.2). The results are presented at few characteristic instants of the cycle – during the compression phase (a), during the fuel injection and the beginning of combustion (b) and in the later part of the combustion process, during the expansion stroke (c). The mixing time values change by two order of magnitudes depending on the angular position during the cycle and the position inside the combustion chamber. The strong changes of mixing time are influenced by the combustion chamber geometry and depend on engine operating parameters such as speed, load and fuel injection strategy.

In the SRM context, the mixing time governs the in-cylinder inhomogeneity of species mass fraction and temperature that have strong influence on the autoignition process, the local rates of heat release and pollutant formation. Hence, the consideration of instantaneous changes of the mixing time during the cycle[1] becomes crucial for an accurate prediction of Diesel engine performance parameters.

1.2.3 Kinetic Models for Diesel Fuel Surrogates

The demand for accurate prediction of engine exhaust emissions promotes the application of detailed reaction kinetics models. These are capable of describing in detail the autoignition, the oxidation and the formation of exhaust emissions from hydrocarbon fuels. Unfortunately, automotive Diesel fuels are complex mixtures of hundreds of components coming from different families of hydrocarbons [107, 162]. Their oxidation process may proceed through thousands of reactions. Because of this complexity, the reaction kinetics-based fuel models are formulated for surrogates, i.e. simpler representations of real fuel blends. The surrogates are comprised of selected species that occur in concentrations devised to ensure combustion characteristics similar to the actual fuels [41].

The foundation of surrogate fuels are chemical reaction mechanisms describing the oxidation process of pure hydrocarbon fuels [123, 142]. These mechanisms can either be used directly as surrogates or are a basis for the derivation of more complex mixtures. Examples of single component Diesel surrogates are straight-chain hydrocarbons such as n-heptane [173], n-decane [63] or

---

[1]In this work henceforth, the term engine cycle or just cycle refers exclusively to the closed part of the engine cycle, unless stated otherwise.

n-dodecane [171]. Multi-component Diesel surrogates are represented by blends of n-decane with 1-methylnaphthalene [60, 136], n-dodecane with m-xylene [60, 75] and several others as reviewed for example in [13, 41, 123].

The possibility to track the formation of different pollutants and reaction pathways in detail is among the advantages of surrogate fuels. Important intermediate or precursory species can be tracked and their impact and contribution to the overall heat release process can be evaluated. A resistance of the fuel to self-ignition, which is one of the key performance parameters of Diesel fuel, can be thoroughly investigated. The surrogates can also be coupled with reaction mechanisms describing NO_x or soot chemistry (see, e.g., [52, 92, 95]) that is important for engine applications.

The use of Diesel surrogates that are derived from the detailed reaction kinetics can be well addressed in engine simulations employing the PDF-based DI-SRM. In this model, chemical reactions are treated exactly. In turn, the 0D formulation enables an effective application of the reaction mechanisms consisting of more than hundred species in still reasonable time frame. With this respect, the use of the DI-SRM and detailed chemistry based Diesel surrogates can be beneficial for investigating various aspects of engine and fuel interactions. This may provide valuable information while designing new fuel blends, developing fuel surrogates for engine simulations [41, 89, 123] or developing engine control strategies.

1.3 Motivation and Objectives

The literature and preliminary works mentioned in previous subsections have indicated the possible benefits of the 0D PDF SRM for the simulation of Diesel engine in-cylinder processes. However, until now there has been rather little attempt to further develop and verify the performance of the DI-SRM variant of the SRM for the simulation of the combustion process of purely Diesel engines. In particular, not much effort has been invested to investigate the possible benefits offered by the crank angle dependent mixing time. This appears crucial for high quality simulations under Diesel conditions using the DI-SRM. The mixing time history must be modelled because 0D models cannot simulate geometry effects on the mixing process. The modelling must consider the impact of changes of engine speed, load and fuel injection strategy on the mixing time, which has also not been investigated so far. Furthermore, not much work has been undertaken to exploit the potential of the DI-SRM for the analysis of fuel effects under Diesel conditions.

The present work aims to fulfil the gaps in the modelling and application rules of the DI-SRM with respect to the simulation of pure Diesel engines and to prove legitimacy of the following thesis:

Representing the occurrence of different flow time scales during the engine cycle improves significantly the overall performance of the DI-SRM for the simulation of combustion, pollutants formation and fuel effects in Diesel engines.

Fulfilling the formulated thesis has been addressed in five specific objectives.

1. Verification of the main modelling concepts and assumptions made for the DI-SRM with respect to the simulation of combustion and emission formation in purely Diesel Engines.

2. Development of crank angle dependent and volume-averaged representative mixing time model for 0D simulations of Diesel engines employing the DI-SRM. Emphasis is on the impact of the modelled mixing time on the prediction of rate of heat release and exhaust emissions such as NO_x, HC and soot.

3. Development of a method for an automate determination of mixing time history for the DI-SRM to make the modelling independent on prior 3D CFD results.

4. Devise a simulation method for the determination of mixing time history before simulations to enable predictive simulations of engine load-speed maps for a variety of operating parameters in Diesel engines.

5. Application of the DI-SRM for the analysis of fuel effects and locality of emissions formation, such as NO_x and soot, in Diesel engines.

1.4 Outline of the Work

The thesis comprises of eight chapters. The content of the chapters is partially based on the previously published works [109, 110, 112–115].

Chapter 1 is a background of the current work. Methods for the simulation of engine in-cylinder processes are outlined. The main challenges faced by 0D simulations of Diesel engines are described. The motivation and objectives of the present study are stated. Chapter 2 provides fundamentals of Diesel combustion and emission formation as a bridge between the physics of a given process and its representation during simulations. The description is limited to the basic information to provide vocabulary for the terms used in this work. Chapter 3 introduces a PDF approach for the simulation of turbulent flows with reactions. The emphasis is on the PDF-based 0D DI-SRM that is the main engine numerical model used in this work. Chapter 4 includes preliminary investigations towards the DI-SRM tailor-made for the simulation of Diesel engines. Modelling assumptions for mixing process, fuel injection and vaporisation are presented and verified. Aspects of computational complexity and accuracy of the results are discussed. Chapter 5 is a main part of the work. It describes the development and validation of the concept of volume-averaged representative mixing time model for the DI-SRM-based simulations of combustion and emission formation in Diesel engines. Applications of the model to engines with single and double fuel injection are presented. Chapter 6 introduces an engineering process for Diesel engine performance studies using the DI-SRM. A procedure for an automate derivation of mixing time history is introduced. Simulation method is presented that enables predictive simulations of engine performance parameters at wide range of engine load and speed changes. Chapter 7 presents several applications of the DI-SRM with the improved modelling of mixing time for the analysis of fuels effects and locality of emission formation under Diesel conditions. Emphasis is on the benefits due to detailed chemistry consideration. Chapter 8 summarises the present work. Conclusions are presented and possible future works are indicated.

Chapter 2

Combustion in Diesel Engines

2.1 Introduction

Fuel injection, mixture formation, combustion and pollutant formation are main in-cylinder processes considered in the simulation of Diesel engines. They are reviewed in the present chapter to introduce terms used in the later parts of this work and to provide a bridge between the physics of a given process and its representation within a numerical model. This assists in the evaluation of the simulation results with respect to the modelling assumptions made. The information presented herein is not aimed at discussing the combustion process in details. These are found in several books and articles [61, 95, 96, 100, 155] that underlay also this chapter.

2.2 Diesel Fuels

Automotive fuels are complex mixtures of hundreds of hydrocarbon components that are derived from crude oil in the refinery processes. The main hydrocarbon groups composing Diesel fuels are usually n-alkanes, iso-alkanes, cycloalkanes and aromatics. The number of carbon atoms in the components varies approximately between 10 and 22, with an average between 14 and 15. Among the representative hydrocarbon components, the iso-alkanes are very slightly branched hydrocarbons containing only one or two methyl groups. The cycloalkanes occur as one ring cyclohexans with multiple alkyl side chains. The aromatics are usually one ring with multiple alkyl side chains.[41, 107, 123, 162]

The primary physical and chemical properties of Diesel fuels that influence the combustion and emission formation processes are density, lower heating value, carbon number range, boiling range, viscosity, sulphur content and composition [107]. These properties depend on the characteristic of the origin of the crude oil and the refinery process that leads to a substantial differences in fuel characteristic between the various world market fuels. These differences underline the need to optimise engine design parameters, such as fuel injection strategy or EGR rate, respectively to the composition of actual fuels to prevent the possible deterioration of engine performance and exhaust emissions if the fuel characteristic is changed.[41]

Most commonly, Diesel fuel is characterised by specifying the cetane number (CN) index that is a measure of the ability of fuel to ignite. The CN index is defined [33] as the volume percent of n-hexadecane (readily igniting, assigned CN=100) in a blend of n-hexadecane and 1-methylnaphthalene[1] (high resistance to ignition, CN=0), which gives the same ignition delay as the test sample. That is, the CN=55 denotes a fuel that in terms of ignition behaves as a mixture of 55% of n-hexadecane and 45% α-methylnaphthalene. A typical value of the CN index for Diesel fuels varies between 40 and 56 [41].

Besides the CN index, the content of sulphur and aromatic components, density and volatility are further controlled fuel properties. They may negatively influence the exhaust aftertreatment devices (sulphurs), contribute to the formation of soot particles (aromatics), affect fuel performance at low temperature (volatility) or interfere with fuel injection systems and mixture formation (density).[162]

2.3 Combustion Process Characterisation

Overall, the combustion process in conventional direct injection Diesel engines is composed of air-fuel mixture formation, ignition and combustion sequence itself. These are the sub-processes governing the rate at which the chemical energy contained in the fuel is released and exhaust emissions are formed. The formation of air-fuel mixture is initiated by fuel injection and it proceeds through fuel spray development, atomisation, vaporisation and air-fuel mixing. The combustion and the mixture formation occur in a cone-like shape fuel spray that is formed at the nozzle exit by the injected liquid fuel jet (Fig. 2.1

---

[1]More recently the heptamethylnonane is used as low CN component instead of 1-methylnaphthalene because of its cost and toxicity [33, 162].

and Fig. 2.2). The characteristic feature of Diesel combustion is concurrent occurrence of mixture formation and combustion. They overlap and interact with each other that makes their modelling a challenging task.[61, 95, 96]

2.3.1 Fuel Injection and Mixture Formation

Fuel injection is a primary process governing the development of a fuel spray. It determines the shape of the spray, the velocity of travel through the combustion chamber and the structure that is described by the fuel droplets size distribution. Furthermore, it influences also the processes occurring inside the spray such as fuel atomisation, vaporisation and mixing.

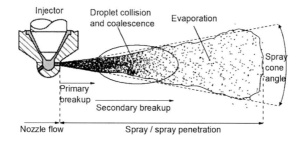

Figure 2.1: Schematic representation of a full-cone Diesel spray with characteristic regimes and lengths (extracted and modified from [14], p.10)

Fuel is injected under high pressure into the combustion chamber through one or usually multi-hole nozzles. High injection pressure generates high relative velocity between the injected liquid fuel and the in-cylinder charge that creates aerodynamic forces. These forces improve the atomisation of the liquid fuel into fine droplets that may vaporise faster. As a result, the mixing intensity and mixture preparation are improved. The first disintegration of the fuel into droplets occurs near the injector nozzle, in the primary breakup regime (Fig. 2.1). Then, in the secondary breakup regime, the already existing droplets further break up into smaller and smaller sized droplets. As the spray penetrates through the combustion chamber with high momentum, air is entrained into the spray. The fuel droplets are heated up due to convective heat transfer and temperature radiation of the hot chamber wall that initiates the evaporation. Subsequently, the vapour fuel mixes with the surrounding gasses and forms the combustible mixture [96]. As these processes proceed, the spray structure disperses taking a cone-like shape as shown in Fig. 2.1.

Mixing and thus the formation of air-fuel mixture are characterised by different time scales during the cycle. On one hand, the intensity of mixture formation is influenced by the fuel-injection parameters such as injection pressure, timing, rate of discharge curve and number of injections. On the other hand, it depends on in-cylinder flows such as swirl, squish and tumble. Initially, the kinetic energy associated with the spray is one order of magnitude higher than that of the in-cylinder flows. Therefore, the injection characteristic is a governing process. Towards the end of injection, as the spray energy dissipates, the influence of the in-cylinder flows becomes noticeable instead. The occurrence of different time scales is a key feature of Diesel combustion that must be respected in engine numerical models.[14, 32, 96]

2.3.2 Ignition and Combustion

Because of the limited time for fuel vaporisation and mixing of the droplets with the surrounding hot cylinder gasses in the spray, there exist fuel rich and lean zones (Fig. 2.2). The mixture is inhomogeneous. An increase of temperature inside the rich zones above approximately 800 K initiates the combustion process that is termed *ignition* or *autoignition*. The crank angle of the ignition timing defines the start of combustion that is controlled by the start of injection. The time gap between the start of injection and the start of combustion denotes the *ignition delay*.

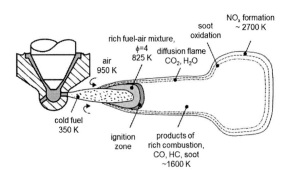

Figure 2.2: Schematic illustration of conventional Diesel spray combustion and emission formation (extracted from [14], p.228)

Among the physical processes occurring during the ignition delay one can list [95] fuel atomisation, vaporisation and mixing of the fuel vapour and air until the ignitable mixture has been created. The chemical processes are pre-reactions (see, e.g., [95, 163]) until the autoignition, which are influenced by the fuel properties such as CN index, pressure, species distribution and temperature inside the combustion chamber.

After the ignition in fuel-rich zones of the spray, the combustion proceeds through (1) *premixed*, (2) *mixing controlled* and (3) *late combustion* phases [61, 95]. Quantitatively, these phases can be analysed using a concept (Fig. 2.3) of the rate of heat release (RoHR) that describes the rate at which the chemical energy contained in the fuel is released by combustion [61]. The premixed combustion is initiated by the ignition process in fuel-rich regimes with local equivalence ratio between 2 and 4. Fuel that is injected during the ignition delay mixes with air and creates easily combustible air-fuel mixture [95]. This mixture burns rapidly, which results in a characteristic peak in the rate of heat release history (Fig. 2.3). The resulting pressure gradients entails engine noise and temperature increase that in turn leads to the formation of NO_x via the thermal NO pathway that is discussed in Section 2.4.

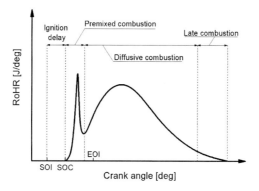

Figure 2.3: Diesel combustion phases (elaborated from [61], p.506)[1]

---

[1]Hereafter in all figures the term deg denotes time in degree on crank angle (CA) basis. Alternative abbreviations are CAD or °CA. Furthermore, time in deg is referred to 0 deg that denotes top dead centre (TDC) during the combustion phase.

Mixing controlled combustion is a dominant mode for Diesel engines. It is initiated after the premixed phase. It continues until the injected fuel has passed through atomisation, vaporisation, mixing with the in-cylinder gasses and burning as diffusion flames. The slowest of these processes and much slower than chemical reactions are vaporisation and mixing. They determine the rate of combustion that is referred to as *mixing controlled combustion*. During the mixing controlled combustion, a majority of the chemical energy contained in the fuel is released, the maximum temperature in the combustion chamber is reached and pollutants are formed.[95]

As the temperature and pressure decrease towards the expansion stroke, the combustion becomes slower and is termed as post-combustion or late combustion. In this combustion mode, chemical reactions slow down. Therefore, chemical kinetics is a dominant mechanism that controls the combustion process. Here, the energy is released due to burning some remaining small amount of fuel that has not burned in the previous combustion phases. Furthermore, the intermediate combustion products, such as unburned hydrocarbons, can be oxidised or a fraction of the energy contained in soot can still be utilised.[61, 95]

In the context of emission formation, the combustion process can be well examined using a concept of equivalence ratio (ϕ) – temperature (T) map such as shown in Fig. 2.4. The ϕ-T maps describe the combined impact of the local in-cylinder temperature and the local availability of fuel and oxygen on the combustion process and the formation of NO_x and soot during the engine cycle. The maps were first introduced in [71] and have become commonly used engineering method [2, 30, 73, 74] to identify engine operating conditions promoting low-emission operation.

Figure 2.4 shows a typical pathway (dots) of the combustion process for conventional Diesel engines. The pathway intersects both NO_x and soot formation regimes. The ignition of the air-fuel mixture is initiated in a rich mixture with local equivalence ratios from the range of 2 to 4 and approximately the temperature of 1500 K. Then, as the temperature increases to about 2000 K, the combustion traverses the regime of soot formation that results from the lack of oxygen for equivalence ratio above 2 (see also Section 2.4). Subsequently, thanks to the mixing process and the temperature increase, the combustion leaves the regime of soot formation and it proceeds towards the stoichiometric condition (ϕ=1). Here, the maximum temperature during the cycle is reached that is associated with the highest rates of NO_x formation. When the equivalence ratio drops below unity, rather insignificant amount of heat is released and the mixing process lowers the temperature of

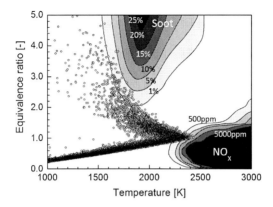

Figure 2.4: Exemplary SRM results (dots) of the in-cylinder local equivalence ratio, ϕ, and temperature, T, plotted over the contours of NO_x and soot formation (adapted from [30]) for the conventional Diesel combustion process

the air-fuel mixture. Overall, the ϕ-T maps illustrate the locality of mixture formation and its inhomogeneity in composition space. They enable the investigation of various combustion concepts with respect to the formation of exhaust emissions.

2.4 Pollutant Formation

2.4.1 Exhaust Gas Composition

In actual engines, the combustion of hydrocarbon fuels is never complete. Besides nitrogen, oxygen, steam and carbon dioxide, the engine exhaust gases contain also pollutants such as NO_x, CO, HC and PM. Furthermore, Diesel fuels include trace amounts of sulphur compounds that are also found in exhaust gases. These pollutants comprise 0.2% by volume of all exhaust gases in Diesel engines [95]. From the point of view of energy losses and their impact on engine processes, such an amount is marginal and can be omitted. However, due to the harmful impact of the pollutants on human health and environment, they are very undesired and hence, restricted by the international law (see, e.g., [40]). Particularly unwanted components of exhaust gasses from Diesel engines are PM and NO_x. They contribute to respiratory issues, lung disorder, acid rains and ground-level ozone [159].

Carbon dioxide is not considered as a pollutant, but due to its contribution to Global Warming it is also subjected to various voluntary agreements such as the Kyoto Protocol [154].

2.4.2 Common Diesel Pollutants

Unburned Hydrocarbons (HC) HC emissions result from the incomplete combustion. The formation of HCs is particularly enhanced during engine cold start and warm-up when the engine and working fluids have not reached operating temperature. Among those most important sources of unburned HCs one can list [61, 95] locally occurring over- or under-mixing of the mixture during the fuel spray development. The unburned HC result also from the occurrence of small zones, such as nozzle sac volume or crevices, where the combustion is limited or does not occur at all. The adsorption of fuel to the oil layer at the combustion chamber wall that prevents its vaporisation and mixing with the air is a next possible source of HC emission. The mixture over-mixing may occur if the small amount of fuel that is injected during the ignition delay mixes rapidly and the local air fuel ratio becomes too lean to initiate the combustion. The under-mixing occurs if the fuel leaves the injector nozzle too slowly. This may lead to the atomisation of over-sized droplets. These are difficult to vaporise and mix with the air sufficiently fast that prevents their full oxidation. These two mechanisms are governed by the mixing process. The composition of the unburned HC depends on the mixing and oxidation processes and is also influenced by the composition of the fuel.

Carbon Monoxide (CO) Similar to the unburned HC, CO also results from the incomplete combustion and is governed by the equivalence ratio. Under lean conditions, when the local equivalence ratio is below unity, CO is oxidised to CO_2. At the near stoichiometric conditions, the oxidation of CO to CO_2 goes via the water shift reaction [95]. In rich mixtures, when the equivalence ratio increases above the stoichiometric value, there is not enough oxygen to burn all the carbons contained in the fuel to CO_2 and the formation of CO increases. Regardless the values of equivalence ratio, significant concentrations of CO are also in the high temperature products of combustion as a result of dissociation. Further contribution to the overall CO concentration level in exhaust gases has its freezing during the expansion stroke due to temperature drop that prevents the oxidation process. Overall however, Diesel engines operate mostly with an excess of oxygen and the formation of CO does not play as important role as in gasoline engines.[61, 95]

Nitrogen Oxides (NO_x) The term NO_x is referred to nitrogen monoxide (NO) and nitrogen dioxide (NO_2). NO is a major component of NO_x emitted by Diesel engines. The contribution of NO_2 in total NO_x in the exhaust gases varies usually between 5% and 30% [61, 86]. Even if the upper limit of NO_2 concentration is reached, it is mostly formed from NO via the reaction with HO_2 that produces NO_2 and OH, i.e. $NO+HO_2 \rightarrow NO_2+OH$ [128, 145, 146]. Therefore, the study of NO_x emitted by Diesel engines is frequently reduced to the analysis of NO formation.

NO is formed via four different mechanisms [163], namely, the thermal NO [82, 172], the prompt NO [42], the NO via N_2O [168] and the fuel-bound NO [66]. With respect to Diesel engines, usually only the first two mechanisms are relevant. The formation of NO via N_2O is more likely to occur in lean premixed combustion such as in gas turbines and sometimes also in gasoline or HCCI engines. In turn, the fuel-bound NO mechanisms is typical in the combustion of coal. As far as the prompt NO is concerned, its contribution to the total NO varies in combustion engines between 5% and 10%. Usually, the prompt NO is relevant for fuel rich conditions in the flame zone and low temperatures since it is initiated at approximately 1000 K. Overall therefore, it is widely accepted that the thermal NO mechanism is the dominant one for combustion engines. It is responsible for 90% to 95% of total NO in the exhaust gases of conventional Diesel engines.[95, 96]

The thermal NO mechanism assumes the formation of NO behind the flame front in the burned gas regime. Originally, the mechanism was postulated by Zeldovich in [172] and later on extended by Lavoie [82]. It is described by three reactions.[61, 95, 163]

$$N_2 + O \rightleftharpoons NO + N \qquad (2.1)$$
$$N + O_2 \rightleftharpoons NO + O \qquad (2.2)$$
$$N + OH \rightleftharpoons NO + H \qquad (2.3)$$

The reaction (2.1) is crucial for the mechanism. It releases nitrogen molecules that are subsequently used in the reactions (2.2–2.3). The reaction (2.1) has high activation energy due to the stable triple bond of the N_2 molecule and hence, it progresses fast only at high temperatures [163]. This is a reason why in Diesel engines most of NO is formed at temperatures above 2000 K (see, e.g., [61]). Besides temperature, local equivalence ratio is the second important parameter governing the formation of NO in Diesel engines. Initially during the cycle, the formation of NO increases due to temperature increase caused by the premixed combustion and resulting pressure increase.

The maximum NO concentration is observed at about stoichiometric mixture and shortly after the maximum pressure peak. A decrease of the in-cylinder temperature due to the expansion stroke and mixing of the gases from hot and cold zones in the combustion chamber entail a decrease of NO formation. However, since the chemical time scales of NO formation and destruction are slower than the physical time scales of temperature changes during the cycle, the reactions do not reach equilibrium conditions. As a consequence, the rate of NO formation is lower than the indication from the equilibrium calculations. In turn, during the expansion stroke and late combustion the backward reactions (2.1–2.3) do not follow fast temperature drop and NO freezes at high concentrations.[61, 95, 163]

Particulate Matter (PM) As the particulate matter is designated the quantity of all substances that are captured by a filter after the exhaust gas has been diluted and cooled down to temperature below 52 °C. Diesel particles consist in the majority of organic components, mainly of pure soot and polycyclic aromatic hydrocarbons (PAH) that are structures composed of aromatic species containing only carbon and hydrogen. These make up of up to 95% of all particles and for this reason PM is frequently referred to as soot particles. Typical size of soot particles from Diesel engines ranges from 10 nm to 1000 nm and the density varies between 1800 kg/m^3 and 2000 kg/m^3.[34, 95]

Details of physical and chemical processes leading to the formation of soot particles have not yet been fully understood. However, it is commonly accepted that soot particles are formed via the formation and growth of PAH species [49]. The process is initiated by fuel pyrolysis and subsequently it involves nucleation/particle inception and processes of coagulation, agglomeration, surface growth and soot oxidation [49, 92, 155]. The formation of soot particles is after [155] illustrated schematically in Fig. 2.5. Fuel pyrolysis leads to the formation of PAH species and acetylene that are soot precursors. Through the nucleation, the first and smallest solid particles are formed from the gas-phase products of the pyrolysis. As the formation of PAH progresses, at a certain point the created structures become no longer arranged on a plane, but represent three-dimensional objectives referred to as particles. These particles undergo chemical and physical processes such as surface growth and oxidation, coagulation and agglomeration. Surface growth is an increase of soot mass through the surface reactions by adding gas-phase species, such as PAH and acetylene, to the particles surface. Coagulation and agglomeration are the processes of formation of larger 3D structures via collision between particles. As the collision proceeds, the initially obtained via coagulation spherical-like structures growth and may form agglomerates [98] that are more chain-like structures. These two processes decrease the number of particles. Oxidation refers to heterogeneous reactions of soot particles and PAH with molecular oxygen and hydroxyl radicals [51, 92]; it reduces the mass of soot. Engine exhaust soot concentration results from the competition between the processes forming and oxidising soot [155].

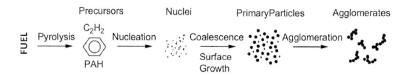

Figure 2.5: Schematic illustration of main chemical and physical processes involved in soot formation (extracted from [155], p.275)

The formation of soot is observed in fuel rich regimes. The unburned fuel or fuel vapour is heated through the mixing with the burned gases with insufficient concentration of oxygen. A significant increase of soot formation is observed in mixtures with equivalence ratio above 2.0 and temperatures range 1700 K – 2000 K (Fig. 2.4). The formation of soot particles strongly depends on temperature and equivalence ratio and less on pressure [7].

2.5 Heat Release by Combustion

In-cylinder pressure is the most commonly measured engine parameter. Combustion, heat transfer to the wall, mass and volume changes are the main processes influencing the crank angle based changes of the in-cylinder pressure during the engine cycle. The thermodynamics analysis allows the evaluation of the contribution of each from these processes to the resulting in-cylinder pressure history. By doing so, the combustion process can be quantified and diagnosed. The rate of heat release is a usual output of the analysis, which is referred to as *heat release analysis*. The history of the rate of heat release describes the rate at which the chemical energy contained in the fuel is released by the combustion process [61]. It correlates to the peak pressure and temperature in the cycle, formation of exhaust emissions and noise generation [23]. Hence, the results of the heat release analysis support the development and validation of numerical models by providing a bridge between the simulated and the experimental data.

2.5.1 Thermodynamic Model

The heat release analysis is based on the first law of thermodynamics applied to an open system. For an internal combustion engine, the term system refers to the in-cylinder charge that consists of air, rest gasses and fuel. For such a system, the first law equation can be written as [53, 61]

$$dU = \delta Q_b - \delta W - \delta Q_{ht} + \sum_{i=1}^{n} h_i dm_i. \tag{2.4}$$

Here, dU is the change in internal energy of the mass contained inside the system, δQ_b represents the chemical heat released during the combustion process, δW is the change in work transfer out of the system and δQ_{ht} is the change in heat transfer out of the system. The term dm_i represents the mass exchange across the system boundary for the ith flow out of n flows that can be due to fuel injection, flows into and out of crevice volumes, blowby and flows through the valves and h_i is the enthalpy associated with that flow. The mass flux term is assumed negative if the mass changes are out of the system.

For the analysis of the combustion process, Eq. (2.4) is usually applied to the in-cylinder charge and during a closed part of the engine cycle, between the inlet valve closure and the exhaust valve opening. Furthermore, it is assumed [22, 23, 46, 53, 78] that combustion is a zero-dimensional and single zone energy release process with uniform distributions of temperature, pressure and composition in the combustion chamber. Thus the content of the cylinder is regarded as a single fluid and its state is expressed by average properties without distinguishing between the burned and the unburned zones. For such a single zone system, Eq. (2.4) can be rewritten to the most commonly used form that with respect to the crank angle increment, $d\varphi$, can be expressed as [22, 53, 61]

$$\frac{dQ_b}{d\varphi} = \frac{\gamma}{\gamma - 1} p \frac{dV}{d\varphi} + \frac{1}{\gamma - 1} V \frac{dp}{d\varphi} + \frac{dQ_{ht}}{d\varphi}. \tag{2.5}$$

Here, Q_b represents the energy released by combustion process and is also referred to as the *apparent gross heat release rate* [61], p is the in-cylinder pressure, V is the cylinder volume, γ is the ratio of specific heats and Q_{ht} is the heat transfer to the wall across the system boundary. It is modelled by the Newton cooling law with the convective heat transfer coefficient that in this work was calculated as in [169].

Thermodynamic properties of the in-cylinder mixture in Eq. (2.5) are represented by the ratio of specific heats (gamma). Gamma is influenced by fuel characteristic, exhaust gas recirculation, charge temperature and air fuel ratio. Frequently, gamma is approximated by a constant value [61] or is varied with the mean temperature of the charge [22] that is calculated from the ideal gas law. In a more accurate approach, the thermodynamics properties of the in-cylinder mixture are calculated by considering fractions of all species composing the mixture, which is possible if detailed chemical reaction mechanisms are used for the description of the oxidation process and emissions formation from a hydrocarbon fuel [87].

By subtracting from Eq. (2.5) the effect of heat transfer to the wall one obtains the *apparent net heat release rate* that is a measure of the energy effectively absorbed by the working fluid. The use of apparent net heat release is frequently preferred over the gross heat release since it eliminates the necessity to calculate the heat transfer to the wall that can be subjected to significant uncertainty. In the later part of this work, the terms *rate of heat release* (RoHR) and *cumulative heat release* (CHR) refer to the apparent net heat release, unless stated otherwise. Furthermore, RoHR traces derived from the measured pressure data are for simplicity referred to as the *experimental* or *measured* RoHR.

2.5.2 Characteristic Parameters

The history of RoHR is the most important output of the heat release analysis. It provides the information about the ignition delay time, the start of combustion (SOC), the end of combustion (EOC) and the amount of heat transferred to the wall. Finally, it determines the total amount of the energy that is released during combustion. All these parameters further complement the information obtained from the in-cylinder pressure data. They enable one to quantify the combustion process.

Following the exemplary RoHR history shown in Fig. 2.6, the SOC and EOC are defined respectively at the CA of minimum and maximum cumulative heat release [61]. The difference between the SOC and the start of injection (SOI) defines the ignition delay. In turn, the difference between the SOC and the EOC determines the combustion duration. Frequently, instead of the actual SOC, the CA of 10% of the released energy (CA_{10}) is used and instead of the actual EOC, the CA of 90% of the released energy (CA_{90}). Crank angles corresponding to these values can be evaluated more accurately than for the actual SOC and EOC. This is true in particular for the EOC that is being approached in an asymptotic manner and therefore it is difficult to precisely determine. Furthermore, the CA of 50% of the released energy (CA_{50}) is also determined that is understood as the combustion centre.

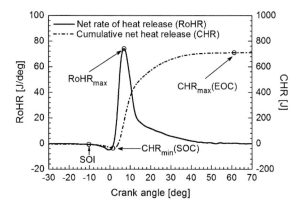

Figure 2.6: Exemplary outputs of the heat release analysis

The heat release analysis allows the distinguishing between the premixed, the diffusive and the late combustion phases in Diesel engines. It enables the determination of the ignition delay time and consequently the combustion phasing. Furthermore, from the RoHR history it is possible to extract some information about the fuel vaporisation process. The process begins after the CA of fuel injection and is revealed by a decrease of the RoHR below the zero level. This negative part of the RoHR history approximates the influence of fuel vaporisation rate. However, it should be noted that this loss of heat is also influenced by the heat transfer to the wall. Furthermore, it can be influenced by the combustion process that might have already started. Hence, it is only a rough estimation of the vaporisation rate. Fuel injection and vaporisation, as well as heat transfer to the wall and combustion overlap partially each other, which prevents an accurate isolation of their individual impacts on the RoHR history.

Chapter 3

Stochastic Modelling of Diesel Engines

3.1 Introduction

The DI-SRM is the primary engine numerical model used in this work. In this chapter, based on previous works [16, 90, 102, 111, 149, 150, 157], an overall concept of the model is presented. The presentation is preceded by the underlying assumptions of the modelling, which result from turbulent flows, reaction kinetics of pollutants formation and the PDF method. This chapter does not intend to discuss these subjects in detail, but it aims to provide a basic information relevant to the DI-SRM-based simulations in this work. A comprehensive treatment of the phenomenology and modelling of turbulent reactive flows can be found in several textbook and articles [47, 59, 69, 76, 80, 85, 91, 118, 126, 158, 163] that also underlay this chapter.

3.2 Turbulent Reactive Flows

3.2.1 Relevance in the Context of Diesel Engines

Turbulent reactive flows are characterised by mutual interactions between the chemical reactions and the turbulent flow. These interactions occur also in Diesel engines and are the primary source of the complexity of Diesel combustion.

Turbulent Flows Turbulence is a feature of the flow that appears if the inertial, buoyancy, centrifugal or other forces overcome the local viscous force that suppress natural disturbances and flow instabilities [58]. The turbulent flows are usually defined by several characteristic features such as [58, 79, 91, 153]: (a) *randomness* – disorder and chaotic character that is reflected in no-repeatability of the properties extracted from the flow such as pressure, temperature or species concentration, (b) *diffusivity* – intensive spreading, mixing and transporting of momentum, heat and species, (c) *energy cascade* – transfer of the turbulent kinetic energy from large eddies towards smaller ones until it dissipates into heat due to viscosity, (d) *dissipation* – increase of the internal energy at the expense of the turbulent kinetic energy, (e) *vorticity and three-dimensionality* since vortex stretching is always three-dimensional. A further feature of turbulent flows are large Reynolds (Re) numbers, continuum phenomenon and occurrence of wide range of time and length scales.

In Diesel engines, the generation of turbulent kinetic energy and turbulence have several primary sources such as tumble, swirl and squish flows. The tumble flow results from the flow through the intake ports and usually it decays during the compression phase. The swirl motion is mainly generated by the geometry of the intake port. The squish flow occurs during the compression phase when the piston approaches top dead centre. Further increase of the kinetic energy is due to fuel injection under high pressure that subsequently penetrates the combustion chamber with a high velocity.[95]

Chemistry The combustion process in Diesel engines proceeds through complex series of thousands of elementary chemical reactions. The complexity of this process results from the composition of Diesel fuels that are mixtures of hundreds of hydrocarbon components. The chemical reactions are highly non-linear functions of composition and temperature. They proceed with different speeds and are characterised by a broad range of time scales.

Turbulence and Chemistry Interactions Chemical reactions occur only if the reactants are mixed on the molecular level and the mixing is caused by the turbulence and eventually molecular diffusion [47, 91, 117]. Hence, the turbulent motion governs the combustion rate. As stated in [118], "the general view is that once a range of different size eddies has developed, strain and shear at the interface between eddies enhance the mixing. During the eddy break-up process and the formation of smaller eddies, strain and shear will increase and thereby steepen the concentration gradients at the interface between reactants, which in turn enhances their molecular interdiffusion." Molecular mixing between the fuel and the oxidiser takes place at the interface between small eddies, bringing the reactants into contact that is necessary to initiate the combustion process. The influence of the chemical reactions on the flow is the result of the heat released by the combustion that increases temperature and may lead to density fluctuations. Buoyancy and gas expansion generate flow instability that enhances the transition to turbulence.[118]

3.2.2 Statistical Modelling and Closure Problem

The difficulties to determine the initial and boundary conditions for real combustion devices and the three-dimensional and chaotic character of the turbulence make the deterministic methods infeasible for the calculation of spatial and temporal evolution of a turbulent flow field with chemical reactions [59, 79]. For these reasons, in engineering applications, statistical methods are used.

The legitimacy of the statistics-based modelling of turbulent flows is reflected in experimental findings. Following [29], "if one measures two times the quantity ϕ in a turbulent flow and at a given location and over the given time interval then one obtains results from these two measurement sets that look quite different. In the same time however, the statistical properties of the measured quantity will be very similar." Thus, even if the properties of a given quantity in a turbulent flow seem highly disorganised and unpredictable, its statistical properties are reproducible.

There are two statistical approaches commonly used [76, 79] for the modelling of turbulent flows with reactions.

1. The method of moments – the properties of the reactive flow are obtained from the transport equations for the statistical moments of the PDF of physical quantities.

2. The PDF method – the properties of the reactive flow are obtained from the transport equations for the PDF of physical quantities.

Both methods introduce the closure problem – new quantities appear in the transport equations that require modelling since the number of unknown variables exceeds the number of available equations.

The overall concept of the method of moments relies on averaging the instantaneous balance equations for the transport of mass, momentum, chemical species and energy to describe the local quantities of the flow [19, 80, 85, 161].

By applying Favre-averaging rules (see, e.g., [79, 161]), the set of transport equations for mass, momentum, chemical species and energy can be, in a generic form, expressed as [118, 124, 161]

$$\frac{\partial(\bar{\rho}\tilde{\phi})}{\partial t} + \nabla \cdot \left(\bar{\rho}\tilde{\phi}\tilde{u}\right) = \nabla \cdot \left(\overline{\Gamma_\phi \nabla \phi}\right) - \nabla \cdot \left(\overline{\bar{\rho}\phi''u''}\right) + \overline{S}_\phi. \tag{3.1}$$

Here, ϕ denotes a generalised flow property, u is the velocity vector, Γ_ϕ is the generalised diffusion coefficient, ρ is the density and S_ϕ is the generalised source or sink term for ϕ. The over-bar denotes time (Reynolds) averaging and tilde mass-weighted (Favre) averaging. The double prime symbol denotes deviation of a given property from the Favre-averaged value. The first term in Eq. (3.1) is the transient term. The second term represents convection. The third and the fourth term are respectively laminar and turbulent diffusion. The fifth term contains all the effects that create or destroy the property ϕ.

The unclosed terms in Eq. (3.1) are $\overline{\bar{\rho}\phi''u''}$ and $\overline{S}_\phi$. In the analysis of turbulent reactive flows, the first from these two terms and for ϕ being species mass fraction (Y) represents in the species conservation equation the species turbulent flux $(\overline{\bar{\rho}Y''u''})$. In turn, the $\overline{S}_\phi$ corresponds to the mean chemical source term that describes the mass rate of production of species i from a unit volume (see $\dot{\omega}$ in Eq. 3.25). Hence overall, these two terms relate respectively to the chemistry and fluid mechanics of turbulent reactive flows [85]. They are subjected to modelling as presented in [47, 118, 124, 166].

The modelling of mean chemical source term is a main problem of the method of moments in turbulent combustion. The problem, after [161], can be illustrated by examination of the production rate of a one step irreversible reaction between fuel (F) and oxidiser (O) that gives (1+S) products (P), i.e. $F + sO \rightarrow (1 + s)P$, for which the fuel mass production rate $(\dot{\omega}_F)$ can be expressed by the Arrhenius law.

$$\dot{\omega}_F = A\rho^2 T^b Y_F Y_O \exp\left(\frac{-T_A}{T}\right) \ , \ T_A = \frac{E_{act}}{R_u} \tag{3.2}$$

Here, T_A is the activation temperature, E_{act} is the activation energy, ρ is the density, R_u is the universal gas constant, T is the mean temperature, A is the pre-exponential frequency factor and Y_F and Y_O are the mass fractions of fuel and oxidant, respectively.

The reaction rate given by Eq. (3.2) is strongly non-linear due to temperature fluctuations caused by the turbulence. As a consequence, the mean reaction rate $(\bar{\omega}_F)$ cannot be expressed as a function of the mean mass fractions, the mean density and the mean temperature. For this reason it requires modelling [118, 124, 161]. The difficulty to express the $\bar{\omega}_F$ as a function of mean properties, particularly temperature, represents the importance of turbulence and chemistry interactions.

Difficulty of modelling the mean chemical source term is avoided in the PDF method. In this approach the chemical source term is treated exactly without modelling. In turn, closure is needed for the mixing process caused by the turbulence.

3.3 Probability Density Function (PDF) Modelling

3.3.1 Background

In the PDF method applied to a turbulent reactive flow, the selected properties of the flow, such as species concentration, temperature or velocity, are represented by the PDF distribution. One can distinguish between the PDF of a single or more variables (joint PDF). The PDF is a dependent variable. It is constructed separately at each location in physical space and time that is denoted as one-point, one-time PDF. Through the mathematical moments, the PDF contains a complete statistical description of the state of the fluid at the separate points in a reacting flow. The shape of the PDF can be presumed or is calculated from a modelled transport equation. The latter method is termed as a transported PDF method and its application relies on the describing the evolution of the one-point and one-time PDF for the selected properties of a reacting flow. The one-point and one-time formulation of the PDF determines main features of the PDF method.[59, 69, 126]

- The chemical source term is treated accurately. This is possible since the separate description of the state of the fluid at each point and time allows for the solution of the chemical reaction rates at these points without approximation.

- The molecular transport needs to be modelled since the information
  contained in the PDF concerns points and times in the flow separately.
  There is no information about what happens between different points in
  the flow; the PDF does not contain the gradient information.

The chemical effects are captured by reaction mechanisms describing the
oxidation process and emission formation from hydrocarbon fuels. The effects
of molecular transport are captured by the modelling of micromixing (see
Section 3.5.5 and Section 5.2).

The velocity-composition joint PDF and the joint composition PDF are two
main variants of the transported PDF method. The velocity-composition
joint PDF is the most complete formulation. For the solution of the flow and
thermochemical properties, it requires modelling only the turbulent molecular
mixing. In turn, in the joint composition PDF, besides the molecular mixing,
the velocity and turbulent diffusivity have to also be modelled separately.
Nonetheless, the composition PDF remains still a valuable method as it
allows one to solve for the chemical reactions without approximation. This is
beneficial in applications oriented at the prediction of exhaust emissions as
for example in IC engines.[126, 127]

If the space effects and consequently the velocity field are excluded from the
analysis, then the resulting composition joint PDF can also be denoted as
0D PDF. Such a formulation of the PDF method is a foundation for the
DI-SRM that is the primary numerical tool used in this work. More detailed
information about the PDF method is found in [59, 126] that underlay the
information presented in this chapter.

3.3.2 Statistical Description

Sample Space and Random Variables

As random variable is considered a variable whose value changes in a stochastic
manner, i.e. its value cannot be determined in advance. With respect to
turbulent flows, examples of random variables can be velocity, enthalpy and
species mass fraction. The value of the random variable cannot be determined
prior to realisation of the flow due to stochastic nature of turbulence. However,
using the PDF formulation, it is possible to ascribe probabilities to its value
being within a certain regime that is realisation (sample-space) of the random
variable. Doing so, the PDF is then considered as the dependent variable.[126,
127]

PDF, CDF and Expectation

Every random variable ϕ has a corresponding sample-space variable ψ. The probability (P) that ϕ is within a regime of ψ is obtained from the cumulative distribution function (CDF) that is denoted by $F_\phi(\psi)$. The CDF represents probability that the random variable ϕ, with a given distribution, will be found at a value less than ψ.

$$F_\phi(\psi) = P(\phi < \psi) \tag{3.3}$$

1. The CDF adopts values from 0 to 1, i.e. $0 \leq F_\phi(\psi) \leq 1$ since for the event $(-\infty < \phi < \infty)$ that is certain one obtains $P(\phi < -\infty) = F_\phi(-\infty) = 0$ and $P(\phi < \infty) = F_\phi(\infty) = 1$.

2. For a given event ϕ, such that $\psi_1 \leq \phi < \psi_2$ and where $\psi_2 > \psi_1$, holds $F_\phi(\psi_2) \geq F_\phi(\psi_1)$. As ψ varies from $-\infty$ to ∞, the $F_\phi(\psi)$ increases from 0 to 1, i.e. $F_\phi(\psi)$ is non-decreasing function of ψ [126].

The derivative of the cumulative distribution function $F_\phi(\psi)$ defines the probability density function $f_\phi(\psi)$ (PDF) of the variable ϕ.

$$f_\phi(\psi) = \frac{d}{d\psi} F_\phi(\psi) \tag{3.4}$$

For an infinitesimal region $d\psi$, Eq.(3.4) can be expressed as $f_\phi(\psi)d\psi = P(\psi \leq \phi < \psi + d\psi)$, i.e. the probability of finding ϕ in a regime $d\psi$ can be obtained from the PDF in that regime. The PDF has three fundamental properties [126].

1. Since the event $(-\infty < \phi < \infty)$ is certain then

$$\int\limits_{-\infty}^{\infty} f_\phi(\psi)d\psi = 1. \tag{3.5}$$

2. Since $F_\phi(\psi)$ is a non-decreasing function of ψ, its derivative $f_\phi(\psi)$ cannot be negative.

$$f_\phi(\psi) \geq 0 \tag{3.6}$$

3. As $F_\phi(\psi)$ tends monotonically to 0 or 1, its derivative $f_\phi(\psi)$ tends to zero, i.e. $f_\phi(-\infty) = f_\phi(\infty) = 0$

If the PDF is known, then the expectation (denoted by angled bracket) of a quantity Q, which is a function of the random variable ϕ, is obtained [126].

$$\langle Q(\phi) \rangle = \int\limits_{-\infty}^{+\infty} Q(\psi) f_\phi(\psi) d\psi \tag{3.7}$$

When applying PDF methods to variable density flows, instead of the PDF the mass density function (MDF) is used as advised in [126]. In composition space, the MDF can be expressed as

$$\mathcal{F}(\psi) = \rho(\psi) f(\psi). \tag{3.8}$$

Here, $\mathcal{F}(\psi)$ is the joint scalar MDF and is regarded as the expected mass density or mass-based discretised PDF.[1]

Discrete Representation of the PDF

Discrete representation of the PDF plays a central role for the solution of the modelled PDF transport equations using the Monte Carlo particle method. The foundation of the discrete PDF is given by the fine grained PDF that denotes the PDF of one realisation of the flow. After [126] it is defined as

$$\langle \delta \left(\psi - \phi \right) \rangle = f_\phi \left(\psi \right). \tag{3.9}$$

Equation (3.9) states that the PDF is the expected value of the Dirac delta function (δ). The Dirac delta function [167] is viewed as the derivative (meaning distribution) of the Heaviside step function H that is defined by $H(x) = 0$ for $x < 0$ and $H(x) = 1$ for $x > 0$. Integrating Eq. (3.9) (see also Eq. 3.4) with respect to ψ gives the CDF.

$$\langle H \left(\psi - \phi \right) \rangle = F_\phi \left(\psi \right) \tag{3.10}$$

Assuming that ϕ represents a single point in the flow at a particular position and time and that from each realisation of the flow different values of ϕ are obtained ($\phi = \phi^{(n)}$), where $n = 1, ..., N$ with N being the number of

---

[1]In this work, the term PDF is used when discussing an overall concept of the PDF method. In turn, the term MDF is used when it is explicitly referred to a mass-based discretised PDF as for example in Section 3.3.3 (Eq.3.20) and Section 3.5.3.

realisations, the ensemble averaged ($\langle \phi \rangle_N$) can be defined as [126]

$$\langle \phi \rangle_N \equiv \frac{1}{N} \sum_{n=1}^{N} \phi^{(n)} \tag{3.11}$$

and the discrete and ensemble averaged of a quantity $Q(\phi)$ as

$$\langle Q(\phi) \rangle_N \equiv \frac{1}{N} \sum_{n=1}^{N} Q\left(\phi^{(n)}\right). \tag{3.12}$$

Based on Eqs. (3.9), (3.10) and (3.11), the ensemble-averaged CDF ($F_{\phi_N}(\psi)$) and PDF ($f_{\phi_N}(\psi)$) for N samples are obtained.

$$F_{\phi_N}(\psi) \equiv \langle H(\psi - \phi) \rangle_N = \frac{1}{N} \sum_{n=1}^{N} H\left(\psi - \phi^{(n)}\right) \tag{3.13}$$

$$f_{\phi_N}(\psi) \equiv \langle \delta(\psi - \phi) \rangle_N = \frac{1}{N} \sum_{n=1}^{N} \delta\left(\psi - \phi^{(n)}\right) \tag{3.14}$$

From Eq. (3.13) and Eq. (3.14) it follows that the PDF of a given random variable can be represented by an ensemble of delta functions with magnitude of $1/N$, where N is the number of samples. The corresponding CDF is represented by an ensemble of step functions (Fig. 3.1).

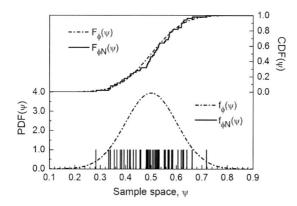

Figure 3.1: PDF and CDF as continuous (dash-dotted lines) and ensemble averaged discrete distributions (solid lines) (elaborated from [126])

Joint PDF

In practical applications there can be more than just one random variable. Besides the information about each variable separately, their joint information is important. The joint CDF and PDF of two variables ϕ and U can after [126]) be defined by analogy to Eqs. (3.4), (3.13) and (3.14). The joint distribution function $F_{U\phi}(V, \psi)$ is given by

$$F_{U\phi}(V, \psi) = P(U < V, \phi < \psi), \tag{3.15}$$

where the event $(U < V, \phi < \psi)$ denotes the probability of finding simultaneously the random variables U and ϕ at values less than V and ψ, respectively. Analogously as for the single distribution function, the joint distribution function is bounded by 0 and 1, i.e. $0 \leq F_{U\phi}(V, \psi) \leq 1$. Furthermore, $F_{U\phi}(-\infty, \psi) = F_{U\phi}(V, -\infty) = 0$, $F_{U\phi}(\infty, \psi) = F_{\phi}(\psi)$ and $F_{U\phi}(V, \infty) = F_U(V)$ and $F_{U\phi}(-\infty, \psi)$ is a non-decreasing function of V and ψ [126]. The derivative of the joint distribution function, $F_{U\phi}(V, \psi)$, defines the joint PDF of U and ϕ in the 2D sample space V-ψ.

$$f_{U\phi}(V, \psi) = \frac{\partial^2 F_{U\phi}(V, \psi)}{\partial V \partial \psi} \tag{3.16}$$

By analogy to Eq. (3.7), the mathematical expectation of a quantity Q, being a function of U and ϕ, can be determined from the joint PDF as

$$\langle Q(U, \phi) \rangle = \int\limits_{-\infty}^{+\infty} \int\limits_{-\infty}^{+\infty} Q(V, \psi) f_{U\phi}(V, \psi) dV \, d\psi. \tag{3.17}$$

Similarly as for the single random variable (Eq. 3.14), the joint PDF can also be represented as an ensemble of Dirac delta functions. As an example [126], if at a given location in sample space, ϕ adopts values $\phi = \phi^{(n)}$ and U adopts values $U = U^{(n)}$, then by summing such events $(n = 1, 2, ..., N)$ the joint PDF $(f_{U\phi N}(V, \psi))$ is obtained as

$$f_{U\phi N}(V, \psi) \equiv \frac{1}{N} \sum_{n=1}^{N} \delta\left(V - U^{(n)}\right) \delta\left(\psi - \phi^{(n)}\right). \tag{3.18}$$

Here, the two-dimensional Dirac delta functions represent unit spikes in the $V - \psi$ sample space. An ensemble of these functions gives a discrete representation of the joint PDF.

3.3.3 Composition Joint PDF Transport Equation

The foundation of the composition joint PDF equation is the transport equation describing the evolution of the composition scalar variables such as mass fraction and enthalpy. After [126] the equation can be expressed as

$$\frac{D\phi_i}{Dt} = -\frac{1}{\rho}\frac{\partial J_\alpha^i}{\partial x_\alpha} + S_i \, , \qquad (3.19)$$

where ϕ_i is the scalar variable, x is the spatial coordinate, α is the summation index, J_α^i is the molecular flux of ϕ_i, S_i is the source term describing the rate of generation of ϕ_i due to chemical reactions [126], where $i=1,\ldots,N_S$ and N_S is the number of scalars. Usually, as scalar variables, ϕ, are considered species mass fraction and enthalpy from which the local properties of the flow with reactions can be determined [59]. The joint PDF of the scalars at the given time and space location (x,t) is denoted by $f_\phi(\psi;x,t)$ and is defined as a probability of a joint event such that $\phi = \psi$. That is, for a set of random composition variables $\phi = (\phi_1, \phi_2, ..., \phi_{N_\phi})$ there is a set of corresponding sample-space variables $\psi = (\psi_1, \psi_2, ..., \psi_{N_\phi})$ such that $(\phi_1 = \psi_1, \phi_2 = \psi_2, ..., \phi_{N_\phi} = \psi_{N_\phi})$.

Expressing the PDF by a MDF (see Eq. 3.8) and using Favre averages, the transport equation for the joint composition MDF that corresponds to Eq. (3.19) can be written as [9, 126, 127]

$$\frac{\partial \mathcal{F}_\phi\left(\psi;x,t\right)}{\partial t} + \frac{\partial\left[\widetilde{U}_\alpha \mathcal{F}_\phi\left(\psi;x,t\right)\right]}{\partial x_\alpha} + \frac{\partial\left[S_i\left(\psi\right)\mathcal{F}_\phi\left(\psi;x,t\right)\right]}{\partial \psi_i} =$$
$$\frac{\partial}{\partial \psi_i}\left[\left\langle\frac{1}{\rho}\frac{\partial J_\alpha^i}{\partial x_\alpha}\middle|\psi\right\rangle \mathcal{F}_\phi\left(\psi;x,t\right)\right] - \frac{\partial\left[\langle u_\alpha''|\psi\rangle\,\mathcal{F}_\phi\left(\psi;x,t\right)\right]}{\partial x_\alpha}. \quad (3.20)$$

Here, the MDF is denoted by $\mathcal{F}_\phi\left(\psi;x,t\right)$, S_i is the source term due to chemical reactions, $\widetilde{U}_\alpha$ and u_α'' are respectively the mean and fluctuating velocities according to Favre averaging. The term J_α^i is the molecular mass flux of species i, where $i=1,\ldots,N_S$ and N_S is the number of species, x is the spatial coordinate and α is the summation index. The first term on the left-hand side of Eq. 3.20 describes the change of the MDF in time. The second term represents the rate of change of the MDF due to convection in physical space with the mean velocity. The third term represents the transport of the MDF in composition space due to chemical reactions. These terms are in closed forms and do not require modelling.

On the contrary, the terms on right-hand side of Eq. 3.20 must be deduced from external models. The first term represents the transport of the MDF in composition space due to molecular mixing. In turn, the second term represents the transport of the MDF in physical space due to turbulent velocity fluctuations. A single point PDF provides no explicit information about the spatial structure and time or length scales [59]. Thus, the gradient field information is also not known [76] that entails modelling of the right-hand side terms of Eq. 3.20. Usually, the mean velocity is modelled by a mean momentum equation that is solved in conjunction with the transported PDF equation [126]. Molecular mixing is modelled by stochastic models (see, e.g., [97, 106, 130] and Section 3.5.5). The transport in the physical space is modelled by gradient-diffusion [127].

The major advantage of the modelled composition joint PDF is revealed in an exact treatment of the source term due to chemical reactions. At every location in space and time, the PDF can be represented by an ensemble of N samples, which are referred to as particles. Each particle is determined by the set of composition variables $\phi = (\phi_1, \phi_2, ..., \phi_{N_\phi})$. For each particle from the ensemble the source term can be determined from the reaction rate expression. As a result, the last term on the left-hand side of Eq.(3.20) is in a closed form and no further modelling is required that is beneficial over other methods in turbulent reactive flows.

3.3.4 Monte Carlo Simulations

The composition joint PDF $f_\phi(\psi; x, t)$ is in general a function of up to $N_\phi+4$ independent variables; N_ϕ of species in the reaction mechanism, three spatial coordinates α and time t. For applications such as the analysis of emission formation in Diesel engines, a reaction mechanism of surrogate fuels can easily contain more than hundreds of species [60, 136, 173]. Consequently the joint PDF is defined in a highly dimensional space and conventional grid-based numerical methods are rather impractical for the solution of such systems. They can only be used for the simplest cases [127]. For this reason, currently almost exclusively the Monte Carlo (MC) particle method, in various variants (see, e.g., [59, 126, 132]), is used [126].

The basic idea of MC simulations [127] is to represent the properties of the flow in the physical domain by a discrete PDF in the computational domain as an ensemble of N notional particles (Fig. 3.2). For a given volume V that contains mass M, each particle represents a mass $\Delta m = M/N$. At a given time t, every n^{th} particle has the composition $\phi^{(n)}(t)$ and position $x^{(n)}(t)$. In

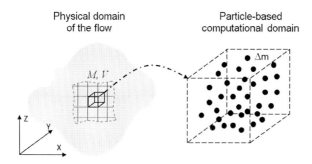

Figure 3.2: Physical domain of the flow and its computational counterpart in the context of Monte Carlo simulations

each time and at any location in the domain, the mean quantities of the flow properties, such as species concentration, can be calculated as an ensemble averages (Eq. 3.11) of the particles in the computational domain. The time evolution of the PDF (or MDF), which describes the stochastic increments of the particles properties, is calculated using one of the available stochastic models (see, e.g., [97, 106, 130]).

3.4 Chemical Kinetics

3.4.1 Reaction Kinetics

Chemical reactions are processes resulting from the collision between molecules if due to these collisions some of the chemical bonds are broken and others are formed, which consequently creates new molecules. Such an elementary process of breaking or forming of a chemical bond is called an elementary reaction. The set of elementary chemical reactions of a given reaction mechanisms can after [117] be written as

$$\sum_{i=1}^{N_S} \nu'_{ij} \, \mathcal{M}_i \; \rightleftharpoons \; \sum_{i=1}^{N_S} \nu''_{ij} \, \mathcal{M}_i \; \text{ for } j = 1, ..., N_R, \tag{3.21}$$

where ν' and ν'' denote respectively the stoichiometric coefficients of reactants and products, $\mathcal{M}_i$ denotes the chemical symbol of species i, N_S is the number of species and N_R the number of reactions.

For each elementary reaction, the law of mass action states that the rate of
formation or depletion of a species is proportional to the product of molar
concentrations of the reactants, each raised to the power of the corresponding
stoichiometric coefficient. Thus, the net reaction rate, r_j, can be defined as

$$r_j = k_{fj} \prod_{i=1}^{N_S} [C_i]^{\nu'_{ij}} - k_{bj} \prod_{i=1}^{N_S} [C_i]^{\nu''_{ij}}, \qquad (3.22)$$

where k_j is the rate constant of jth reaction and the subscripts f and b stand
for forward and backward reactions, respectively. The net reaction rate of jth
reaction is denoted by r_j. The term $[C_i]$ denotes the molar concentration of ith
species out of N_S species and ν'_i and ν''_i denote the stoichiometric coefficients for
reactants and products, respectively. In Eq. (3.22) the reaction rate constant
k is a constant specific for each elementary reaction and is expressed by the
modified Arrhenius law [163].

$$k = AT^b \exp \left(-\frac{E_a}{R_u T} \right) \qquad (3.23)$$

Here, A is the pre-exponential factor, b is the temperature exponent, E_a is
the activation energy, R_u and T are the universal gas constant and absolute
temperature, respectively.

During reactions mass is conserved. The molar net rate of formation of a
species in a multi-step reaction mechanism can be obtained by summation
over all net reaction rates with their relevant stoichiometric coefficients.

$$\omega_i = \sum_{j}^{N_R} \nu_{ij} r_j \qquad (3.24)$$

Here, ω_i denotes the molar net rate of formation of species i in the jth reaction,
ν_{ij} is the net stoichiometric coefficient $(\nu_{ij} = \nu''_i - \nu'_i)$ and r_j is the net reaction
rate of the jth reaction out of N_R reactions. Using the molar net rate of
formation, ω_i, of a species i and the molar mass of that species, W_i, the mass
rate of production, $\dot{\omega}_i$, from a unit volume, V, is obtained.

$$\dot{\omega}_i = \omega_i W_i V \qquad (3.25)$$

The mass rate of production of species given by Eq. (3.25) corresponds to the
source term $\overline{S}_\phi$ in Eq. (3.1) that represents the effects of chemical reactions.
In the PDF calculations of turbulent reactive flows, this term is given in a
closed form and does not require modelling.

3.4.2 Analysis of Reaction Mechanisms

A reaction mechanism is a sequence of elementary chemical reactions that is followed by an overall combustion process. The reaction mechanism describes what happens during each stage of the transformation from the combustion reactants to products. It shows which chemical bonds are broken or formed, what the sequence is of these processes and what the reaction paths are that species follow. This information allows one to determine the rate of species production or depletion that is needed for the solution of Eq. (3.1) and Eq. (3.20).

Following [62, 80], there are four categories of reaction mechanisms: (1) detailed, (2) reduced, (3) semi-global and (4) single-step. They are characterised by a different level of complexity and hence, are tailored for different applications. For the simulations oriented on the prediction of exhaust emissions the first two groups are relevant. They are briefly outlined here, following the descriptions in [80, 163].

Detailed Mechanisms This group denotes mechanisms that contain possibly a complete set of reaction kinetics describing the oxidation process of a hydrocarbon fuel. Such complete detailed mechanisms that would be valid for all operating conditions and different applications are rarely available [163]. The reason is the complexity of the combustion process that for commonly used hydrocarbon fuels may easily follow a thousand of elementary reactions and involve hundreds of species [163]. For such large systems it is difficult to determine all possible intermediate reactions. As a consequence, as detailed mechanisms are considered these when no simplification has been carried out on purpose and when researchers have done their best to consider all existing reactions [80]. For the solution of practical engineering problems the application of the detailed mechanisms is still not straight yet, mainly, because of the high computational cost, but also because the detailed mechanism provides frequently more information than is necessary.

Reduced Mechanisms To overcome the difficulties with the detailed reaction mechanisms, they are simplified and reduced to a small size and computationally less demanding models. The reduced models contain only these species and reaction paths that are important for a given problem under consideration. As a consequence they are valid only for a given range of conditions that was targeted during the reduction process. Such reduced mechanisms are specific problem tailored mechanisms and are relevant for practical applications such as in IC engines.

Mechanisms reduction is performed by applying different mathematical tools such as sensitivity analysis, reaction flow analysis and species life time analysis [3]. *Sensitivity analysis* determines species that are the most and least important with regards to the chosen control parameters. *Reaction flow analysis* measures the flow of atoms that occurs between species through the reaction pathways. Consequently, the importance of each pathway is determined. *Species life analysis* investigates the reaction mechanism in terms of the range of time scales present in the system. It determines short time scale species that are often associated with the local equilibrium process and can be removed from the mechanism.

A separate issue is a validity of the reduced mechanism under real operating conditions. Usually reaction mechanisms are validated against experimental data from ideal experimental cases such as a perfectly stirred reactor or a plug flow reactor, where flow and thermodynamics parameters can be controlled. In actual applications however, the flow and thermodynamics conditions may vary significantly. Such situations occur in IC engines, where there is a continuous interaction between chemistry and turbulent flow. For this reason a care must still be taken while applying the reduced models for the analysis of real engineering problems.

3.4.3 Soot Modelling

Numerical models for the description of soot formation and oxidation are divided into empirical, phenomenological or semi-empirical and models with detailed chemistry. The models with detailed chemistry are the most comprehensive. They attempt to describe elementary physical and chemical process of soot formation starting from fuel pyrolysis through complex reaction kinetics of the gas-phase chemistry up to the formation of solid particles and their oxidation.[72, 155]

Detailed Kinetic Soot Model The detailed soot model is used in this work.[1] The model was developed originally in [94] and thoroughly described in [92]. Its different applications are discussed in [7]. A detailed presentation of the model is beyond the scope of this work. The description below is confined to the basic concept of the model for the readability of this work, in

---

[1]The description of the model is derived in part from an article [115] published in Combustion Science and Technology on 30 September 2014, available online: http://wwww.tandfonline.com/DOI: 10.1080/00102202.2014.935213.

particular Section 7.4. Main processes distinguished by the model with respect to soot formation are briefly outlined based on the description presented in [7, 92, 94].

Particle Inception The coagulation of two polycyclic aromatic hydrocarbon (PAH) structures forming a primary and 3D particle. The model allows the inception of the first smallest particle through the coagulation between two PAH structures, each containing four aromatic species at least [137].

Condensation The coagulation of a PAH larger than four aromatic rings coagulates on the surface of a soot particle, which leads to soot mass growth [92, 94].

Coagulation and Agglomeration The formation of larger 3D structures via collision between particles that decreases the number of particles; as the collision proceeds the initially obtained via coagulation spherical-like structures growth and may form agglomerates [98] that are more chain-like structures.

Surface Growth An increase of soot mass through the surface reactions that is modelled following the hydrogen-abstraction-acetylene-addition (HACA) mechanism [51] with a separate ring closure (HACARC) [92].

Fragmentation The abstraction of an acetylene molecule from the soot surface [92]; a contrary process to surface growth.

Oxidation Heterogeneous reactions of soot particles with molecular oxygen and hydroxyl radicals [51, 92]; it reduces the mass of soot particles.

Method of Moments As a result of different physical and chemical processes affecting soot formation and oxidation, the soot particles occur in various size classes. Hence, the prediction of soot formation requires the calculation of the soot particle size distribution function (PSDF) or its main features. In this work, the method of statistical moments, proposed in [50, 51], has been used to characterise the soot PSDF. The statistical moments are defined as

$$M_r = \sum_{i=1}^{\infty} i^r N_i \qquad r = 0, 1, ..., \infty. \tag{3.26}$$

Here, M_r is the rth soot moment, N_i is the number density of a particle of size class i and the mass of particle is expressed as $m_i = i \cdot m_1$, where m_1 is the mass of two carbon atoms.

The moments due to the soot source terms, particle inception, condensation, coagulation and heterogeneous surface reactions are calculated as in [9, 92]. The first two moments are used. Moment M_0 corresponds to the total particle number density,

$$M_0 = \sum_{i=1}^{\infty} N_i = N, \tag{3.27}$$

regardless the size of particles. In turn, moment M_1 corresponds to the total soot mass, regardless the number of particles.

$$M_1 = \sum_{i=1}^{\infty} i N_i = Y_s \frac{\rho}{m_1} = f_v \frac{\rho_s}{m_1} \tag{3.28}$$

Here, ρ is the density, m_1 is the mass of the smallest soot monomer unit, ρ_s is the soot density, Y_S is the soot mass fraction and f_v is the soot volume fraction.

By defining transport equations for $N_i/(\rho \cdot N_A)$ and applying Eq. (3.26), a general transport equation for the soot moments can be derived (see, e.g., [92, 122]).

$$\rho \frac{\partial M_r^*}{\partial t} + \rho v_\alpha \frac{\partial M_r^*}{\partial x_\alpha} - \frac{\partial}{\partial x_\alpha} \left(\rho D_{p,1} \frac{\partial M_{r-\frac{2+\theta}{3}}^*}{\partial x_\alpha} \right) - \frac{\partial}{\partial x_\alpha} \left(0.55 \rho v \frac{1}{T} \frac{\partial T}{\partial x_\alpha} M_r^* \right) = \omega_r. \tag{3.29}$$

Here, x is the spatial coordinate, α is the summation index, $M_r^* = M_r/(\rho \cdot N_A)$, v is the flow velocity, $D_{p,1}$ is the diffusion coefficient and θ is a constant used to consider the effect of the fractal dimension of the soot particles. The source terms due to particle inception, condensation, coagulation, surface growth and oxidation are denoted overall by ω_r. Details of the closure of the source terms are presented in [92].

The soot model, based on the moments of the soot PSDF, enters the DI-SRM through Eq. (3.30) that defines the joint vector $\phi(t)$ of the composition random variables. The moments are mixed in phase space in the same way as the other random variables such as enthalpy and species mass fraction.

The first two moments provide information about the total number of soot particles and soot volume fraction. Usually, this information sufficiently characterises the soot PSDF in practical applications.

3.5 Direct Injection Stochastic Reactor Model (DI-SRM)

3.5.1 Background

The SRM for 0D simulations of internal combustion engines has been proposed in [77, 90] to study HCCI combustion. The model evolved from the deterministic reactor models such as a plug flow reactor model (PFR) and a perfectly stirred reactor model (PSR). These reactors are commonly used for the analysis of interactions between thermodynamics, reaction kinetics and fluid mechanics in chemical engineering [45].

The PFR is a cylindrical pipe with reactants flowing down the pipe. The flow is considered to be a steady-state, plug flow type and frictionless. There is no mixing and variation of velocity and concentration in the axial direction. The PSR is a vessel with intake and outlet ducts and an impeller inside for mixing. The mass flow through the reactor is considered to be steady-state. The content of the reactor is assumed to be perfectly homogeneous. There is no time or position dependence of the temperature, concentration or reaction rate inside the reactor.[45, 158]

The PFR and PSR do not consider the effects of mixing due to turbulence-chemistry interactions that occur in reacting flows. This is the main drawback of these models that results from the simplified assumption that the mixture is perfectly homogeneous. To overcome this shortcoming, in [76], randomness has been introduced into the mixing and heat transfer processes. The obtained stochastic variants of the PRF and PSR models are denoted as partially stirred plug flow reactor (PaSPFR) and partially stirred reactor (PaSR), respectively. This modelling was further extended in [77, 90] to enable the simulation of engine in-cylinder process. The model obtained is referred to as the stochastic reactor model (SRM) for engine applications.

In the next sections, a basic description of the SRM is given based on [9, 16, 54, 55, 76, 76, 87, 88, 90, 101, 111, 126, 134, 157]. The description focuses on the main features and assumptions of a variant of the SRM for the simulation of direct injection engines (DI-SRM).

3.5.2 Overall Formulation

The overall concept of the DI-SRM for the simulation of combustion and emission formation in Diesel engines is based on the PDF approach for the simulation of turbulent flows with reactions. In particular, it relies on the particle-based formulation of the composition joint MDF transport equation that is introduced in Section 3.3.

The 0D DI-SRM considers gas inside the cylinder as an ensemble of notional particles as schematically shown in Fig. 3.3. The particles represent a one-point and one-time PDF for a set of scalar variables. The scalar variables (ϕ) correspond to species mass fraction, $Y_i(t)$, and enthalpy, $h(t)$, from which the temperature is determined, i.e. $\phi(t) = (Y_1, \ldots, Y_{N_S}, h; t)$, where $i = 1, \ldots, N_S$ and N_S denotes the number of species in the reaction mechanism. The length of the vector ϕ has a dimension of $N_S + 1$ that corresponds to the size of the reaction mechanism, N_S, and the enthalpy, h, of a single particle.[1] Each particle of a given mass is fully determined by the set of scalar variables and it represents a point in gase-phase for species mass fraction and enthalpy.

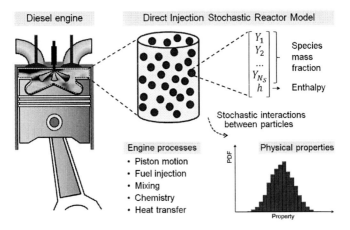

Figure 3.3: Concept of the PDF-based 0D DI-SRM for Diesel engines

---

[1] If the method of moments is used for the prediction of soot formation, the length of the vector ϕ is increased by the number of the calculated moments (see Section 3.4.3).

The scalars, enthalpy and species concentration, are considered random variables that can vary within the cylinder. They determine the composition of the gas mixture and are described with probabilities using the PDF. The in-cylinder mixture is thus represented by a PDF in gas-phase and the particles realise the distributions. Overall, rather than PDF, the MDF is used for variable density flows [126] that for equally weighted particles can be considered as a mass-based discretised PDF.

The particles are subjected to physical and chemical processes relevant for the simulation of Diesel engines such as piston movement, chemical reactions, heat transfer, mixing and direct fuel injection. Furthermore, the particles can also mix with each other and exchange heat with the cylinder walls. The space effects on these processes is not considered hence, 0D modelling framework. The local quantities, such as temperature or species mass fraction that in actual engines are distributed in the space, in the DI-SRM their distributions are determined by the MDF.

3.5.3 MDF Transport Equation

The joint composition MDF transport equation (Eq. 3.20) is a foundation of the DI-SRM. It provides a solution for the random scalar variables such as enthalpy $h(t)$ and species mass fraction $Y_i(t)$ that enable calculating the release of chemical energy due to combustion. The joint vector $\phi(t)$ of the random scalar variables is defined as

$$\phi(t) = (Y_1, \ldots, Y_{N_S}, h; t) = (\phi_1, \ldots, \phi_{N_S}, \phi_{N_S+1}; t), \qquad (3.30)$$

where N_S is the number of species in the reaction mechanisms and ϕ_{N_S+1} corresponds to enthalpy, h. The vector of the random scalar variables has a corresponding joint scalar MDF vector that is expressed as

$$\mathcal{F}_\phi(\psi; t) = \mathcal{F}_\phi(\psi_1, \ldots, \psi_{N_S}, \psi_{N_S+1}; t). \qquad (3.31)$$

Here, $\psi_1, \ldots, \psi_{N_S}, \psi_{N_S+1}$ is the vector denoting the realisation of the random variables $\phi_1, \ldots, \phi_{N_S}, \phi_{N_S+1}$. Furthermore, it is assumed that probabilities of all scalar variables are independent of position – statistical homogeneity. Thus, the MDF does not vary spatially within the cylinder.

$$\frac{\partial}{\partial x_\alpha} \mathcal{F}_\phi(\psi; x, t) = 0 , \qquad \alpha = 1, 2, 3 \qquad (3.32)$$

As a consequence, the transport of the MDF in physical space due to convection with the mean velocity and velocity fluctuations is omitted. Hence, the time evolution of the MDF, which is given by Eq. (3.20), is reduced to (see, e.g., [9, 76])

$$\frac{\partial}{\partial t}\mathcal{F}_\phi(\psi;t) + \frac{\partial}{\partial \psi_i}(Q_i(\psi)\mathcal{F}_\phi(\psi;t)) = \frac{\partial}{\partial \psi_i}\left(\left\langle \frac{1}{\rho}\frac{\partial J_\alpha^i}{\partial x_\alpha}\bigg| \psi \right\rangle \mathcal{F}_\phi(\psi;t)\right). \qquad (3.33)$$

Here, the MDF is denoted by $\mathcal{F}_\phi(\psi;t)$. The initial conditions are given as $\mathcal{F}_\phi(\psi;0) = \mathcal{F}_0(\psi)$ and $\mathcal{F}_0(\psi)$ represents the initial distribution at time $t = 0$. The first term on left-hand side describes the change of the MDF in time. The second term represents the transport of the MDF in composition space. The term Q_i is a source/sink operator. It describes engine in-cylinder process that is presented in next subsection. The right-hand side term of Eq. (3.33) represents the transport of the MDF in composition space due to molecular mixing, where ρ is the density, J_α^i is the diffusive flux vector for i, x is the spatial coordinate and α is the summation index. This term requires modelling since the MDF does not consider the space effect and the gradients cannot be determined. An exemplary method of modelling this term is presented in Section 3.5.5.

Equation (3.33) describes the PaSPFR model [9, 76] that underlies the SRM for engine applications [90]. The equation is solved sequentially for each considered engine in-cylinder process using an operator splitting loop method [87] that is introduced in Section 3.5.6.

3.5.4 Diesel In-Cylinder Processes

For Diesel engines, the operator Q_i in Eq. (3.33) represents the change of the MDF due to chemical reactions, convective heat loss, volume change due to piston movement and fuel injection. The impact of these processes on the MDF is calculated based on the species and energy transport equations (see, e.g., [19, 80, 85, 161]). These equations with respect to the operator Q_i can be expressed as [15, 87, 157]

$$Q_i = \frac{W_i}{\rho}\omega_i + \frac{\dot{m}_f}{m}\left(Y_{i,f} - Y_i\right), \qquad (3.34)$$

$$Q_{N_S+1} = \frac{dp}{dt} + \frac{h_g A}{V}(T - T_w) - \sum_{i=1}^{N_s} H_i\omega_i + \frac{\dot{m}_f}{V}Y_{i,f}\left(h_{i,f} - h_i\right). \qquad (3.35)$$

Equation (3.34) and Eq. (3.35) describe the conservation of species mass fraction and energy in terms of specific enthalpy. They are formulated with respect to Diesel engine processes. In these equations, h_g is Woschni's heat transfer coefficient, T_w is the cylinder wall temperature, A is the heat transfer area, i=1,..., N_S, where N_S stands for the number of species i. Subscript f denotes fuel. H is the molar specific enthalpy and is given by $H = h_i W_i$, where h_i is the specific enthalpy of species and W_i denotes the molar mass of species i. The molar net rate of formation ω_i of species i is calculated from Eq. (3.24). The variables p, T and V correspond to in-cylinder pressure, gas temperature and instantaneous volume, respectively.

In Eq. (3.34), the two terms on the right hand side represent respectively the impact of chemical reactions and fuel injection on the mixture composition. In turn, the terms on the right hand side of Eq. (3.35) represent respectively the influence of pressure changes, heat transfer to the wall, chemical reactions and fuel injection on the in-cylinder mixture.

Chemical Reactions and Heat Release Chemical reactions are represented in Eq. (3.34) and Eq. (3.35) by the first and the third term, respectively that contain the molar net rate of formation ω_i of species i. These terms are solved at a given time step for each particle. As a result, the chemical composition of the particles at the new time step is changed. The heat release due to chemical reactions is calculated based on the known mass of each particle and the particle enthalpy before and after a chemical reaction. The cumulative chemical heat release is obtained as a sum of the heat being released from each particle in the ensemble and in each time step during the cycle.[87]

Piston Movement The instantaneous volume change due to piston movement is calculated based on the known engine geometry. For a slider-crank mechanism it can be calculated as in [61].

$$V(\varphi) = V_c + \frac{\pi}{4}B^2(l + a - a\cos(\varphi) + \sqrt{l^2 - a^2\cos^2(\varphi)}) \qquad (3.36)$$

Here, φ corresponds to time in crank angle degree, B is the cylinder bore, l is the connecting-rod length, a is the ratio of the crank radius and V_c is the clearance volume. Depending on the design of the crankshaft mechanism, Eq. (3.36) can be modified to include also the effects of piston pin and crank offsets (see, e.g., [87]).

Heat Transfer In Eq. (3.35), heat transfer to the wall is calculated deterministically using the Newton cooling law with Woschni's convective heat transfer coefficient [169]. To consider the influence of wall temperature and overall mixture inhomogeneity in temperature on the heat transfer rate, the stochastic fluctuations are introduced into the Woschni's heat transfer coefficient. How the heat is transferred over the particles is determined on a stochastic basis as presented in detail in [16, 87]. Loosely, in this algorithm, on randomly chosen particles a temperature jump is performed with reference to the wall temperature. The magnitude of the jump is a modelled parameter C_h that is termed as a stochastic heat transfer constant. High value of C_h decreases the number of particles participating in the heat transfer event, whereas low value increases it. Consequently, the evolution of the particles due to the stochastic heat transfer mimics the effects of the heat transfer process in actual engines.

Fuel Injection The injection process is modelled by introducing into the ensemble of existing particles (air and rest gases) new particles that represent fuel. The process of adding fuel particles is governed by the fuel injection rate profile. Adding fuel particles changes the total mass inside the cylinder. Furthermore, the mass fractions and temperatures of the particles in the ensemble also change because the injected fuel has a certain composition and temperature. The effects of adding new particles is represented by the last terms on the right hand sides of Eq. (3.34) and Eq. (3.35).[87, 157]

Mixing The modelling of molecular mixing involves prescribing the evolution of stochastic particles in composition space such that they mimic the change of the composition of a fluid parcel in actual turbulent reactive flows [130]. In practise, the modelling requires providing the closure for the molecular mixing term in the joint composition PDF transport equation (see [59, 126]). The realisation of the mixing process is of primary importance with respect to the modelling of mixing time that is discussed in Chapter 5 of this work. For this reason, mixing is discussed in more detail in next subsection.

By solving Eq. (3.33) with the help of submodels for Diesel engine processes, such as piston movement, fuel injection, mixing, chemical reactions and heat transfer to the wall, one obtains information about the PDF of species concentrations and temperature inside the combustion chamber. This information allows the determination of engine performance parameters of practical importance such as in-cylinder pressure, apparent rate of heat release and pollutant formation.

The in-cylinder pressure, p, at each angular position during the cycle is calculated from the ideal gas law.

$$p(\varphi) = \langle \rho(\varphi) \rangle \frac{R \langle T \rangle}{\langle W \rangle} \tag{3.37}$$

Here, φ denotes time in crank angle degree, $\langle T \rangle$ and $\langle W \rangle$ are the mean temperature and molar mass, respectively and $\langle \rho(t) \rangle$ is the mean density that after [90] is calculated as $\langle \rho(\varphi) \rangle = m \cdot V(\varphi)^{-1}$. The apparent heat release is calculated based on the known in-cylinder pressure history obtained from Eq. 3.37 and the thermodynamics properties of the in-cylinder mixture as presented in Section 2.5. Pollutant formation is calculated based on the reaction mechanisms (see Section 3.4.1) and using the information about species concentration from all particles in the ensemble at each time step.

3.5.5 Mixing Modelling

Several models have been proposed for the description of turbulent mixing as reviewed in [37, 97, 106, 130]. Some of them have also been used within the SRM and the PaSPFR models. For example, the interaction by exchange with the mean (IEM) [35] was used in the SRM [90] to investigate the influence of mixture inhomogeneity caused by the thermal boundary layer adjacent to the cylinder walls on the HCCI combustion process. In [9] the IEM model, the coalescence-dispersion (CD) model due to Curl [28] (denoted also as Curl's model) and the Binomial-Langevin model [160] were used within the PaSPFR to simulate a furnace black process.

In the present work the Curl's mixing model has been used. Loosely, in this model, the mixing process is realised as follows. At a given discrete local sub-time step $(t+\delta t)$, out of the ensemble of particles, two particles (m,n) are selected on a random basis. The particles mix with each other and the values of their scalar properties (ϕ) are replaced by the common mean that is calculated [106, 125, 126] from the values before the mixing at the previous time step (see Eq. (3.38) and Fig. 3.4).

$$\phi^{(n)}(t + \delta t) = \phi^{(m)}(t + \delta t) = \frac{1}{2} \left(\phi^{(n)}(t) + \phi^{(m)}(t) \right) \tag{3.38}$$

The scalar properties mixed, ϕ, are species mass fractions and specific species enthalpy from which the temperature is determined. The selection of a pair of particles and their mixing are repeated sequentially with the locally calculated sub-time step, δt, (details of the implementation in [87]). During each global simulation time step, Δt, the number of particle pairs selected for mixing, $N_{P_{mix}}$, depends on the total number of particles, N_p, and the decay rate of the scalar variance, ω_ϕ, that after [106] is expressed as $N_{Pmix} = \omega_\phi \Delta t N_P$. The number of sub-time steps, δt, which corresponds to the number of mixing events between the randomly chosen particle pairs, is inversely proportional to the mixing time that is an input to the mixing model. The mixing of particle pairs continue as long as the sum of the sub-time steps, δt, is smaller than the global time step, Δt, in the operator split loop (Fig. 3.5).

The evolution of particle properties due to CD model changes discontinuously in time (Fig. 3.4) yielding non-continuous PDF that does not relax to the Gaussian distribution. The model does not fulfil the locality criterion imposed on mixing models (see, e.g., [80, 151]). For example, it allows, through the random selection of particles, mixing a very rich particle with a very lean one, yielding two new particles that do not necessary ignite. This might correspond to the unrealistic mixing across the reaction zone.[59, 80]

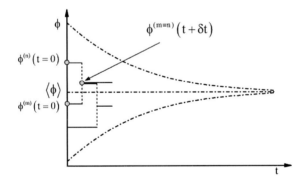

Figure 3.4: Particles trajectory in ϕ space according to the CD model [126]

The common feature of the mentioned mixing models is the necessity to know the mixing time before simulations. This is one of the modelling challenges for the DI-SRM and is discussed in Chapter 4 and Chapter 5.

3.5.6 Numerical Solution

The Monte Carlo particle method, outlined in Section 3.3.4, with the operator splitting technique as presented for example in [87, 126] is employed to obtain the solution for Eq. (3.33). In this method, the solution during each time step is obtained by a separate calculation of the impact of volume changes, fuel injection, mixing, chemistry and heat transfer on the MDF (Eq. 3.33). The evolution of the MDF due to these processes is calculated sequentially as illustrated in Fig. 3.5.

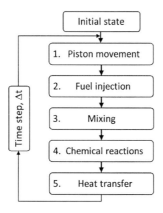

Figure 3.5: Operator splitting loop algorithm [87]

The initial state of the MDF (see Section 3.3.2) is expressed as an ensemble of delta functions corresponding to the number of particles with given values of initial temperature and species mass fractions. Each of the processes 1 to 5 in Fig. 3.5 contributes to the MDF distributions regardless of the other ones. The MDF is updated with the result from each process with the assumption they occur at a constant pressure. The engine in-cylinder performance parameters at each time step are obtained by "summing up" the individual results of the steps 1 to 5 in the operator splitting loop.

Solving Eq. (3.33) sequentially simplifies numerical efforts, but it also introduces an error when compared to the direct solution. After each sequential step the thermodynamics conditions of the particles, particularly pressure, change due to the calculated processes. To ensure that different steps in the operator splitting loop are all performed at the same conditions, the pressure change must be corrected. This is done [87] by compressing adiabatically all particles in the system, such that the sum of the volume of all particles is equal to the total cylinder volume at the given CA. After the correction, the temperature and density of all particles are updated according to the new pressure value. The pressure correction (details in [87, 134]) is performed after each sequential step shown in Fig. 3.5.

Chapter 4

DI-SRM Tailor-Made for Diesel Engines

4.1 Introduction

The DI-SRM is a general purpose model for the simulation of combustion in IC engines with direct fuel injection. Respectively to the treatment of in-cylinder processes, such as fuel injection, vaporisation and mixing, the model can be employed for the analysis of various combustion modes. This chapter introduces preliminary assumptions and modelling concepts that are necessary to tailor the DI-SRM specifically for the simulation of conventional Diesel engines with direct fuel injection. In the first part of the chapter main indications and assumptions for using crank angle dependent mixing time in the DI-SRM are introduced. The description provides a foundation for a detailed discussion of this subject that is presented in Chapter 5. Subsequently, the treatment of the fuel injection and vaporisation processes is discussed. Next, the influence of model parameters on the overall performance and accuracy of the results from the DI-SRM is analysed. Finally, the effects of reaction mechanisms are discussed.

4.2 Mixing Time for the DI-SRM

With respect to Diesel engines, mixing time can be viewed as an inverse of the frequency at which air, fuel and rest gases mix with each other during the cycle. Thus, it represents time scales of the micromixing.

Mixing results from the diffusion processes. In turbulent flows it can be divided into molecular (or scalar) mixing and mixing due to velocity fluctuations (or just turbulent mixing as referred to in this work). These two transport processes may occur at different time and length scales and can be viewed as a random movement of molecules and eddies, respectively.[76, 91, 126]

Both turbulent flow mixing and molecular mixing and their characteristic time scales play a central role in the PDF-based DI-SRM. The molecular mixing term in the joint composition MDF equation (see Section 3.5.3) occurs in an unclosed form and must be modelled. The closure is provided via micromixing models (see, e.g., [37, 59, 97, 106, 126]) that simulate the composition change of stochastic particles in the gas-phase for the DI-SRM. The turbulent mixing time is a crucial modelled parameter of the micromixing models. It governs the frequency of mixing between the particles. Hence, it influences the in-cylinder inhomogeneity for local values of equivalence ratio and temperature. These in turn, govern the autoignition, local rates of heat release and pollutant formation in Diesel engines.

The occurrence of different time and length scales during the Diesel cycle [61, 146] is a basis for the use of a crank angle dependent mixing time in the DI-SRM. This is in contrast to the simulation of HCCI combustion, where single value mixing time was frequently used [3, 17, 90, 102, 150]. Crank angle (or time) dependency of the mixing time in Diesel engines is confirmed by 3D CFD calculations in Fig. 1.1 that shows also space dependency of the mixing time.

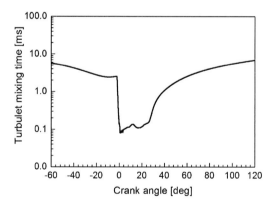

Figure 4.1: Example of turbulent mixing time history from 3D CFD calculations of a Diesel engine operated with single fuel injection

The space dependency of mixing time can be reduced into 0D domain as it is presented in Fig. 4.1. In that figure, the history of mixing time was calculated from the ratio between the turbulent kinetic energy (k) and its dissipation (ϵ) (see also Section 5.2.1). The values of k and ϵ were obtained from 3D CFD computations that were carried out for the whole combustion chamber and for different angular positions during the cycle. Thus the resulting history represents time dependency of the mixing time with respect to crank angle changes. The results show a substantial variation of mixing time scales during the cycle. Especially noticeable is a sudden increase of the mixing intensity, thus short time scales, at approximately -3 °CA ATDC that corresponds to the time of fuel injection. The high intensity is kept until the injection has finished. Subsequently, it decays exponentially towards the exhaust valve opening (EVO). The presented history of mixing time is characteristic for engines operated with single fuel injection. It is also a basis for the modelling presented in Chapter 5.

4.3 Direct Fuel Injection and Vaporisation

The formulation of the SRM assumes that the notional particles represent the gas-phase of the in-cylinder mixture of actual engines. Hence, with respect to the simulation of direct injection Diesel engines, it is necessary to model the changes of the mixture state in the gas-phase due to fuel injection and vaporisation. In the DI-SRM used in this work, the mass of fuel injected is assumed to vaporise instantaneously at the moment of injection. This, in practise, implies the fuel injection rate curve to represent the vaporisation rate [87, 157]. A practical consequence of this assumption is that SOI becomes the start of vaporisation (SOV) and EOI estimates the end of vaporisation (EOV). Furthermore, the angular position of the SOV determines the beginning of the intensive mixing due to fuel injection (see, e.g., mixing time in Fig. 4.1, around the TDC and shortly after). This simplified modelling is application tailored. Depending on the conditions inside the cylinder, mainly temperature at the moment of fuel injection, the rate of fuel injection may not necessarily reasonably mimic the rate of fuel vaporisation. To ensure possibly accurate representation of the vaporisation process, in this work as a rule of thumb, the SOV being derived from the SOI is a calibrated parameter. The purpose of the calibration is to retard the input fuel injection rate profile to ensure an accurate prediction of the SOC timing. The concept is shown schematically in Fig. 4.2.

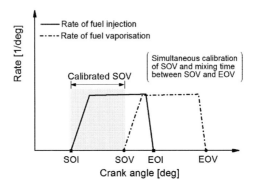

Figure 4.2: Concept of the simulation of fuel vaporisation rate

How far the calibrated SOV must be delayed with reference to the actual SOI depends strongly on the mixing intensity that controls the vaporisation process. Thus in fact, the prediction of the SOC is governed by simultaneous calibration of the SOV and mixing time. The combined impact of the SOV and mixing time is discussed separately in Chapter 5. This section focuses only on the effect of vaporisation timing and rate. For this reason and to simplify the description, a constant value mixing time was used during the calibration of the vaporisation rate.

Parameter studies were performed to verify the proposed modelling of the vaporisation rate. Calculations were carried using one representative fuel injection rate profile in the DI-SRM. Five different angular positions of the SOV were considered. With reference to the position of SOI (-2 °CA ATDC), the SOV was retarded with 1 °CA interval and between -1 °CA ATDC and 4 °CA ATDC. A constant value mixing time was used during the fuel injection/vaporisation phase of the mixing process and equal to 0.2 ms. The resulting vaporisation rate profiles are presented in Fig. 4.3.

In Fig. 4.3 the solid line denotes the reference vaporisation rate profile. It was estimated from the heat release analysis of the experimental in-cylinder pressure data (see also Section 2.5). The profile starts at SOI=-2 °CA ATDC and it proceeds until approximately 4.5 °CA ATDC. The modelled vaporisation rates are denoted by discontinuous lines. They were extracted from the heat release analysis of the simulated in-cylinder pressure data. The zero-level corresponds to the value of RoHR for the reference vaporisation rate (solid line).

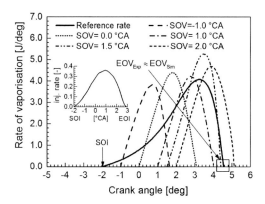

Figure 4.3: Heat loss due to fuel vaporisation. Solid line: experimental-based history. Dashed lines: DI-SRM computations for different SOV positions

The modelled fuel vaporisation histories are shorter than the reference history. Their higher rates ensure that the same mass of fuel has been vaporised as for the reference rate profile. The biggest discrepancy between the reference and the modelled traces is observed during the initial phase of vaporisation, where the characteristic long tail vaporisation is not captured by the modelled profiles. This results from the assumption that the modelled vaporisation rate is a derivative of the fuel injection rate that does not contain the initial long tail. To improve this inaccuracy, besides calibrating the SOV, one may also calibrate the duration of the input rate profile. This can be a necessity if for example the considered fuel blend contains slowly vaporising components and consequently there is a substantial difference between the duration of fuel vaporisation between different components of the fuel.

Alternatively, the fuel vaporisation rate obtained from the heat release analysis of the in-cylinder pressure data could be applied. However, such a determination of the vaporisation rate is also only an estimation. The vaporisation and the combustion process overlap partially each other and it is difficult to define precisely the end of vaporisation. Further impact is due to the performance of the heat release model applied. Thus, this method of estimating the rate of fuel vaporisation may also require calibration. Furthermore, the method requires measured in-cylinder pressure as an input thus, its application is limited to already existing data.

In practise, the main demand imposed on the modelled for the DI-SRM fuel vaporisation is to ensure an accurate prediction of the begin and the initial increase of the combustion rate. This demand can effectively be captured by the proposed simplified modelling. For SOV=1.5 °CA the end of the modelled vaporisation rate matches well the end of the reference, experimental-based profile (Fig. 4.3). As a consequence, the simulated rate of heat release and start of combustion match well the experimental counterparts as it is presented in Fig. 4.4.

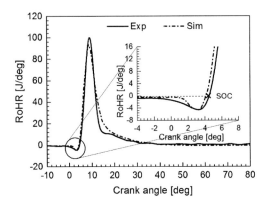

Figure 4.4: Rates of heat release and the heat loss due to vaporisation

Good agreement between the simulated and the experimental rates of heat release in the regime where fuel vaporisation takes place (zoom in Fig. 4.4) indicates the proposed modelling of the vaporisation process be sufficiently accurate and relevant for the present investigations. Further benefit of the proposed method is simplicity of the use. The method is based on the fuel injection data that is an engine design parameter and usually known before simulations. Thus, it is possible to model the effects of fuel vaporisation for operating points for which no experimental in-cylinder pressure data is available. The calibration is carried out in an automate manner that is introduced in Section 6.3.

4.4 Model Parameter Studies

4.4.1 Design Parameters

Overall, a setup of the DI-SRM is determined by a set of numerical and physical parameters. The numerical parameters are the number of stochastic particles, the number of consecutive cycles, the calculation time step in the operator splitting loop and the amplitude of the stochastic heat transfer. The parameters influence the stability and accuracy of the results. The modelled physical parameters are fuel injection with vaporisation and mixing. They influence physical and chemical in-cylinder processes simulated by the DI-SRM

Mean-Cycle Results The DI-SRM simulates heat transfer, mixing and fuel injection as stochastic processes (see Chapter 3). As a result, there are cycle-to-cycle differences between the simulated parameters such as pressure or exhaust emissions. To ensure the simulation results are repeatable and comparable to the experimental data, calculations must be carried out with sufficiently small time step, high number of particles and over several consecutive cycles. The mean properties obtained from the cycle-to-cycle results are considered as the representative outputs for the comparison with the measured data.

Number of Particles (N_P) The number of particles determines the discretisation of the in-cylinder content. It has an impact on the simulated local inhomogeneity of the in-cylinder mixture that in turn influences the calculated engine exhaust emissions such as NO_x, CO, HC and soot. Increasing the number of particles improves the quality of the results. They are smoother and lower cyclic variation is observed. On the other hand, the larger the number of particles the higher the computational cost of simulations.

The negative effect of increasing the N_P on computational cost is partially avoided by using the *particles clustering* method. The method was introduced in [116] and is also used in this work. Loosely, in this method an algorithm is employed to search for the particles of similar properties. Similarities between particles can be based on species concentrations and/or a number of different variables such as enthalpy of formation or equivalence ratio [87]. Such similar particles are then clustered into bigger particle clusters. By doing so, the effective number of computed particles (clusters) is reduced and the overall computational cost of simulations is decreased.

Number of Consecutive Cycles (N_C) The number of consecutive cycles, from which mean values of engine performance parameters are calculated, influences the results in a similar manner as the number of particles. The higher the value of N_C, the more stable and representative are the averaged results and the higher the computational cost. Simulating more cycles and taking their average corresponds to post-processing of the experimental data from steady-state measurements, where averaging is also performed because of the cycle-to-cycle variation (see, e.g., [61]).

Operator Splitting Time Step (Δt) The operator splitting time step refers to the time interval, on a CA basis, of separate calculation of chemical reactions, convective heat loss, piston movement, fuel injection and mixing (see also Section 3.5.6). Smaller Δt increases the accuracy of the simulations by diminishing the negative effect of separate calculation of these processes, which in actual engines occur simultaneously and also interact with each other. On the other hand, smaller Δt increases the cost of computations.

Amplitude of the Stochastic Heat Transfer (C_h) The constant C_h determines how much heat is transferred over different particles (see Section 3.5.4). The heat transferred over a particle affects the formation of NO_x and the oxidation of HC, CO and soot from that particle. However, for engines and operating points considered in this work, the preliminary investigations have indicated rather small sensitivity (Table 4.1) of the calculated NO_x, unburned HC and CO on C_h values.

Table 4.1: Effect of the C_h constant on engine exhaust emissions

Emissions [-]	Stochastic heat transfer constant (C_h) [-]				
	5	15	30	50	90
NO_x	1.02	1.00	0.97	0.98	0.99
HC	0.90	1.00	0.96	0.96	0.97
CO	0.99	1.00	0.98	1.01	1.00

Results normalised to C_h=15 and for N_P=1000, N_C=35, and Δt=0.5 °CA.

The differences are around the uncertainty of the results due to the stochastic nature of the DI-SRM. The influence is small, especially when compared to the influence of the modelled mixing time (discussed in Chapter 5). For this reason, the constant C_h was kept constant and equal 15 (advised in [87]) for all calculations in this work.

Fuel Vaporisation and Mixing Time These are key modelled physical parameters for the DI-SRM. They are discussed separately in Section 4.2, Section 4.3 and Chapter 5.

4.4.2 Accuracy of the Results

A sensitivity study was performed to evaluate the influence of the numerical parameters (N_P, N_C and Δt) on computational cost of simulations and accuracy of the prediction of in-cylinder pressure, rate of heat release and exhaust emissions such as NO_x, HC, CO and CO_2. First, the effects of the numerical parameters were investigated separately for each parameter. After that, cross-check tests were performed by changing simultaneously more than just one parameter. Calculations were performed using the 121-species n-heptane mechanism from [173] that is outlined in Section 4.5.

Accuracy of the simulated results was determined for in-cylinder quantities that were calculated as mean values from multi-cycle runs. Calculations were carried out for several values of N_P, N_C and Δt. To consider the effects of stochastic processes in the DI-SRM (see Section 3.5), the results were collected from ten subsequent calculations for each combination of N_P, N_C and Δt. From the obtained population of results, standard deviation was calculated for each considered in-cylinder quantity. The value of standard deviation is denoted as relative tolerance (in the unit of percent) of the data simulated by the DI-SRM. To consider the effects of the number of probes taken for the calculation of mean values, confidence intervals were additionally determined (see, e.g., [108]) for mean quantities.

Accuracy of the results is analysed based on the results from Case 2 that is characterised in Table B.1 of Appendix B. Preliminary calculations indicated that for $N_P \leq 100$ and $\Delta t \geq 1.0$ °CA the simulated exhaust emissions and in-cylinder pressure have a noisy character and high standard deviations of cycle-to-cycle results. On the other hand, for $N_P \geq 2000$ and $\Delta t \leq 0.1$ °CA no visible improvement of the results was observed.

Maximum in-cylinder pressure (p_{max}) and the concentration of unburned HC at EVO are two of the most sensitive parameters to the changes of N_P, N_C and Δt. If the demanded accuracy of the results has been achieved for HC and p_{max}, the demands assigned to NO_x, CO and CO_2 are usually also fulfilled. For this reason the results presented in this section focus on p_{max} and HC.

Figure 4.5 presents accuracy of the simulated maximum in-cylinder pressure and HC concentrations at EVO. The results were obtained at different combinations of N_P and N_C and for Δt equal 1.0 °CA and 0.5 °CA. Simulations with Δt=0.1 °CA did not show a significant improvement of the accuracy when compared to these with $\Delta t = 0.5$ °CA, but were occupied by higher cost of computations.

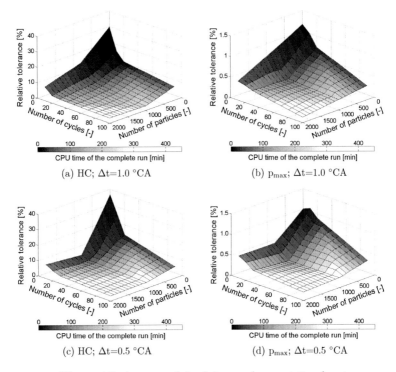

(a) HC; Δt=1.0 °CA

(b) p_{max}; Δt=1.0 °CA

(c) HC; Δt=0.5 °CA

(d) p_{max}; Δt=0.5 °CA

Figure 4.5: Accuracy of simulations and computational cost

Overall, regimes of the smallest tolerances (highest accuracy) of the calculated p_{max} and HC are approximately bounded by N_C>30 and N_P>500 for Δt=1.0 °CA, and N_C>20 and N_P>200 for Δt=0.5 °CA. In these regimes, the relative tolerances for the simulated HC and p_{max} are respectively below 6% and

0.6%, but are occupied by high computational cost. Complete run[1] takes approximately three hours. For higher values of N_C and N_P, the CPU time may increase up to more than seven hours. Calculations were performed in parallel on 32 CPUs of the computer cluster equipped with AMD Opteron 2378 @2.4 GHz processors from the year 2008.

Figure 4.6 shows the influence of the number of CPU units on the computational cost of simulating a single engine cycle. The most noticeable speed up is observed between 1 CPU and 8 CPUs. Above 32 CPUs there is no more visible improvement. Thus, for the selected N_P, N_C and Δt, the use of 16 CPUs or 32 CPUs seem to be the best choice. It ensures a reasonable trade-off between the gained speed-up and CPU loading.

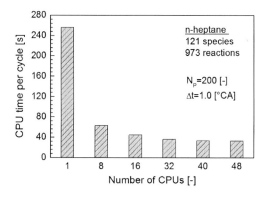

Figure 4.6: The impact of number of CPUs on computational cost

Based on the calculations carried out, three configurations of numerical parameters have been proposed. Their choice was driven by the trade-off between accuracy of the results and CPU time. The allowed uncertainty window for the prediction of unburned HC, NO_x, CO, CO_2 and p_{max} was defined based on the tolerances of the corresponding measured data. Thus,

---

[1]In Fig. 4.5 and also later in the text, the term complete run (or full run) refers to the sum of all consecutively calculated cycles. Due to stochastic processes in the DI-SRM, there are cycle-to-cycle variations in the calculated quantities. Hence, for the comparison with the experimental data, the simulated engine parameters are used as averages from multi-cycle runs. Therefore, in Fig. 4.5, presenting the CPU time for a complete run instead of for just one cycle (single calculation process) is in this case more relevant.

9% uncertainty was assumed for HC, 1% for p_{max} and 4% for NO_x, CO and CO_2. Table 4.2 summarises the performance of the selected configurations of numerical parameters. Besides HC and p_{max}, also tolerances for NO_x, CO and CO_2 are presented.

Table 4.2: Computational cost and tolerance for the calculated mean values of p_{max}, NO_x, CO, HC and CO_2 using the 121-species n-heptane mechanism

Setup	Settings	Tolerance [%]					CPU time (32 CPUs)	
		HC	NO_x	CO	CO_2	p_{max}	Full run [min]	1 cycle [s]
A	N_P=200 [-] N_C=100 [-] Δt=1.0 [°CA]	8.9	3.1	3.3	0.1	0.34	59	36
B	N_P=200 [-] N_C=30 [-] Δt=1.0 [°CA]	13.7	4.4	6.6	0.1	0.85	18	36
C	N_P=1000 [-] N_C=30 [-] Δt=0.5 [°CA]	5.3	1.7	1.5	0.04	0.20	113	228

Setup-A This configuration has successfully been used in several different applications of the DI-SRM having regard to simulations of different engines and operating points. It ensures a stable solution for the in-cylinder pressure, rate of heat release and exhaust emissions. Using the 121-species n-heptane mechanism, the complete run takes approximately 1 hour and the tolerance-window of the simulated HC (8.9%) is below the assumed limit (9%). The accuracy of simulated HC and p_{max}, expressed as relative standard deviations, are $\pm$ 7.5% and $\pm$ 0.27%, respectively, when referred to a 95% confidence interval.

Setup-B This is considered as a simplified version of Setup-A. It was defined for the simulations targeted at the prediction of in-cylinder pressure and rate of heat release. The N_C has been decreased to 30, reducing the CPU time of the full run down to 18 minutes.

Setup-C If more accurate results are demanded, then it is advised to decrease the Δt and increase the N_P. Higher discretisation of the solution domain gives more accurate results, but it requires approximately two times more CPU time when referred to Setup-A. If the local in-cylinder processes, such as mixing or soot formation are to be studied, then it may be reasonable to use N_P=2000 to have even more finer discretisation of the in-cylinder mixture.

The parameter studies presented have not considered the influence of the model parameters on the accuracy of soot prediction. However, recently performed investigations [48] indicate the soot results be similarly sensitive to the changes of numerical parameters as the HC emission.

The investigations presented in this subsection do not define a single setup for all numerical parameters. Instead, they provide a database of solutions from which N_C, N_P and Δt can be defined respectively to the simulation targets.

- For most applications it is reasonable to use N_P between 200 and 2000, Δt between 0.5 °CA and 1.0 °CA and N_C between 15 and 100.

- The Setup-A, with N_C=100, N_P=200 and Δt=1.0 °CA, is considered as basic in this work. It ensures a reasonable trade-off between the accuracy of the results and computational cost in predicting the in-cylinder pressure and exhaust emissions.

- For more detailed investigations, focusing on the locality of the in-cylinder processes, such as mixing or soot formation, the Setup-C is advised.

Based on the investigations carried out, the modelled numerical parameters of the DI-SRM, such as N_P, N_C, Δt and C_h, can be presumed before simulations. The fuel vaporisation rate is determined automatically during the calibration of mixing time. Hence, mixing time history is the only modelled physical parameter of the DI-SRM for the simulation of Diesel engines.

4.5 Aspects of Chemical Complexity

Because of the complexity of detailed reaction mechanisms [13, 41, 123], their reduced versions are usually preferred to represent the performance of actual fuels in industrial applications. These reduced models are applications tailored and hence, are valid for a range of operating conditions that was considered during the reduction process. As a result, the reduced mechanism for a given fuel molecule may perform differently when the operating conditions have changed. Therefore, besides N_P, N_C and Δt, a reaction mechanism also influences the overall performance of the DI-SRM and the computational cost.

4.5.1 n-Heptane as Diesel Surrogate

Among Diesel surrogates, pure n-heptane deserves special attention. It is a single component surrogate that is frequently used in practical applications to IC engines (see, e.g., [84, 120, 144, 152]). The frequent use of n-heptane, though there exist other and more complex multicomponent surrogates [13, 41, 123], results from its similarity to real Diesel fuels in ignition behaviour, which is one of the key parameters influencing Diesel combustion. Further advantage is a relatively broad and well documented database of reaction mechanisms [41] of different complexity and size and validated over a broad range of operating conditions. The main drawbacks are higher than in real Diesel volatility and omitting the effects of aromatics, cycloalkanes and iso-alkane components on autoignition that also impacts the prediction of pollutant formation. The drawback due to differences in the physical properties affecting the atomisation and vaporisation can be diminished by taking them from heavier hydrocarbons such as n-decane or n-dodecane. By doing so, only the chemistry is modelled with n-heptane. In comparison to the multicomponent surrogates, another benefit of n-heptane is usually lower demand for CPU time that results from a simpler structure of the mechanisms.

The present work is targeted at the improvement of the overall simulation process and performance of the DI-SRM for the simulation of Diesel engines and not on the surrogate fuels assessment or their improvement. Therefore, the mentioned deficiencies of n-heptane as Diesel surrogate are of secondary importance.

4.5.2 Reaction Mechanism Effects

Two reaction mechanisms for n-heptane have been used in this work. Both mechanisms include the thermal model of NO formation proposed in [82, 172]. The rate constants have been extracted from [141].

- The smaller mechanism, extracted from [156], was used in applications oriented at low computational cost during the development of a simulation process for the DI-SRM. The mechanism contains 28 species and 58 forward and backward reactions. It has been obtained by subtracting the iso-octane pathways from the primary reference fuel (PRF) model. Later in the text the mechanism is referred to as *the 28-species mechanism*.

- The larger mechanism, taken from [173], was used as a basis for the simulation oriented at the prediction of exhaust emissions. The mechanism contains 121 species and 973 backward and forward reactions. Later in the text the mechanism is referred to as *the 121-species mechanism.*

In Fig. 4.7 and Fig. 4.8 are presented the ignition delay times predicted by both reaction mechanisms. The ignition delay has been selected as a tracking parameter because of its decisive impact on the combustion timing in Diesel engines. The simulated and the experimental data are presented for equivalence ratios 0.5 and 2.0 at pressures 13.5 bar and 40 bar.

Under lean conditions (Fig. 4.7) both mechanisms similarly predict the ignition delay times and with high accuracy when compared to the experimental data.

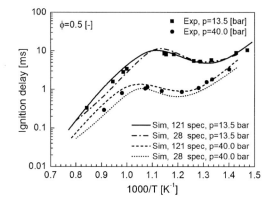

Figure 4.7: Computed and measured ignition delay times for n-heptane at ϕ=0.5. Experimental data from [44] versus computations in a constant volume reactor for the 121-species mechanism [173] and for the 28-species mechanism [156]

Under rich conditions (ϕ=2.0, Fig. 4.8) the results deviate from each other. The highest discrepancy between the two mechanisms is observed in the negative temperature coefficient (NTC) regime. Only the larger mechanism follows the measured points with high accuracy over the whole temperature range. Up to approximately 1000 K, the computed ignition delay times from the 28-species mechanism are much longer than the corresponding experimental data and the results from the bigger mechanisms.

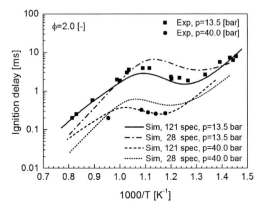

Figure 4.8: Computed and measured ignition delay times for n-heptane at $\phi=2.0$. Experimental data from [44] versus computations in a constant volume reactor for the 121-species mechanism [173] and for the 28-species mechanism [156]

The difference in the predicted ignition delay times by both mechanisms is reflected in the histories of the in-cylinder pressure and mean temperature presented in Fig. 4.9. Calculations were performed using the same settings of the numerical parameters and history of the mixing time in the DI-SRM. Thus, the observed differences result from the differences in chemical ignition delay times for the considered two reaction mechanisms.

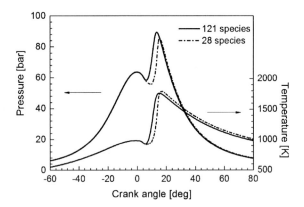

Figure 4.9: Simulated in-cylinder pressure and temperature histories using two different reaction mechanisms and the same setup of the DI-SRM

The small mechanism ignites later in the cycle. This is attributed to longer ignition delay times that are observed in the NTC regime, in particular for rich mixtures (see Fig. 4.8). This behaviour is crucial for fuel autoignition that is initiated at locally rich mixtures, where the shortest ignition delay times are found. In consequence, the application of the smaller mechanism delays the start of combustion and decreases the maximum peak of the in-cylinder pressure. On the other hand, it increases the in-cylinder pressure gradient after the start of combustion that in turn increases the maximum peak of the mean in-cylinder temperature and the temperature at exhaust valve opening.

Despite the differences between the performance of both mechanisms, their applications presented in Chapter 5 and Chapter 6 indicate that both models can simulate reasonably well the in-cylinder pressure and rate of heat release of n-heptane and Diesel fuelled engines. The possible inaccuracy of the reaction mechanism is in some extent compensated during the calibration of the mixing time. The ignition process is governed by the Damköhler number that describes the ratio between the flow time (associated with mixing) and the chemical time [79] (see also Section 5.2.2). In this context, the mixing time can be considered as reaction mechanism tailored.

In contrast, engine exhaust emissions, particularly HC, are simulated less accurately by the smaller mechanism for the same configuration of the DI-SRM parameters (N_P, N_C and Δt). The accuracy of exhaust HC prediction for the 28-species mechanism may rise up to 30% in comparison to 5.3% for the 121-species mechanism. Higher tolerance and thus lower stability of the calculated HC is for this particular mechanism attributed to its development process. The original PRF model [156], from which the 28-species n-heptane scheme was extracted, was developed to study low-temperature combustion in HCCI engines. It contains rather simple high-temperature mechanism that is important for the simulation of Diesel engines. Mixing time calibration cannot compensate for this drawback.

The reduction of computational cost is a main benefit of the small mechanism. It is relevant for the applications targeted at the prediction of in-cylinder pressure and rate of heat release. With regards to exhaust HC prediction, it is reasonable to assume the smaller mechanism be relevant to simulate trends rather than absolute values.

Figure 4.10 compares computational cost depending on the size of the mechanism. Calculations were carried out in parallel on 32 CPUs (2.4 GHz processors from the year 2008). The 28-species model is approximately 5 times faster than the 121-species model. Low demand for CPU time makes the small mechanism useful during simulation methods development, initial investigations and tests, where rather than the absolute values of the simulated performance parameters the trends are important. For this reason the 28-species mechanism has also been used during the development of a simulation process for the DI-SRM, which is presented in Chapter 6.

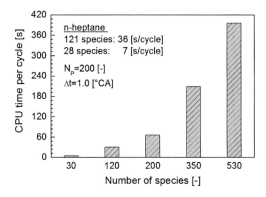

Figure 4.10: The impact of number of species in the mechanism on computational cost of simulations

Chapter 5

Mixing Time Modelling in the DI-SRM Context

5.1 Introduction

The modelling of mixing time is a main focus of this work and is addressed in this chapter. The mixing time is a key modelled parameter of the DI-SRM for the simulation of Diesel engines. The first part of the chapter is devoted to basic information about turbulent flow, mixing and their interactions with chemistry. Thereafter, based on the information from Section 1.2.2 and Section 4.2, a concept of crank angle dependent and volume-averaged representative mixing time model is introduced. The modelled history of mixing time is validated against 3D CFD calculations. The performance of the model is investigated based on the simulation of Diesel engines operated with single and double fuel injection.

5.2 Turbulence and Mixing

5.2.1 Time and Length Scales

The flow processes of Diesel engines are turbulent that among others manifests by randomness, diffusivity, mixing, dissipation and three-dimensional fluctuations of velocity. Among these features, mixing is the one that enables increasing the overall rate of reactions and hence, improving the efficiency of combustion.

Mixing results from the diffusion processes. It is divided into molecular mixing and turbulent flow mixing that may occur at different length and time scales. The largest time and length scales inside the cylinder are bounded by the combustion chamber geometry and the smallest by molecular mixing.

Turbulent Mixing Time (τ_t) is often seen as an inverse of the turbulence intensity caused by the velocity fluctuations (u'). It describes the time for the turnover of the largest structures in the flow that are defined by the integral length scale (l_I). This is expressed by

$$\tau_t = \frac{l_I}{u'}. \tag{5.1}$$

Furthermore, the turbulent mixing time represents also the time taken to transfer the kinetic energy (k) from the large scales l_I to small scales, where due to viscosity it is dissipated into heat. Thus, using Favre averaged kinetic energy and its dissipation (ϵ), the turbulent mixing time can be expressed as [80, 118]

$$\tau_t = \frac{\tilde{k}}{\tilde{\epsilon}}. \tag{5.2}$$

Molecular Mixing Time (τ_ϕ) represents the decay time scale of scalar fluctuations. It describes mixing between the fuel and the oxidiser at the molecular level, where chemical reactions take place. The molecular mixing corresponds to the scalar dissipation rate (χ) that can be expressed as [80, 118]

$$\tau_\phi = \frac{\widetilde{\psi''^2}}{\tilde{\chi}}. \tag{5.3}$$

Here, ψ denotes a scalar quantity (e.g. species mass fraction or mixture fraction) and $\tilde{\chi}$ is the mean scalar dissipation rate. Here, both these quantities are expressed as Favre averages.

Providing a relation between the scalar dissipation rate and the turbulent mixing time is one of the key issues in the modelling of a non-premixed turbulent flow with reactions [80, 118].

5.2.2 Mixing Time Effects

The importance of mixing time is revealed in the influence on flame stability and the occurrence of autoignition. These processes can be analysed using the S-shaped graph of temperature (T) versus the Damköhler number (Da). The Damköhler number is a non-dimensional number that relates the characteristic time associated with turbulent mixing (τ_t) and the characteristic time associated with chemical reactions (τ_{ch}) [79].

$$Da = \frac{\tau_t}{\tau_{ch}} \tag{5.4}$$

The influence of mixing time on temperature can also be related to the inverse scalar dissipation rate, χ, noting that the scalar dissipation rate at the stoichiometric mixture fraction, χ_{st}, represents the inverse of a characteristic diffusion time.[12, 80, 118]

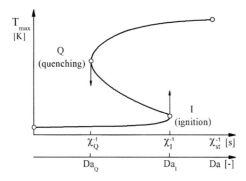

Figure 5.1: S-shaped curve of maximum diffusion flame temperature versus the Damköhler number (adopted from [80], p.146)

In Fig. 5.1, the upper branch of the S-shaped curve corresponds to highly reactive and burning state. The middle part of the diagram, between ignition (I) and quenching (Q), represents an unstable regime that is physically unrealistic. The lower branch represents the state of initially lower reactivity up to the ignition point. This branch is relevant for the description of autoignition process in Diesel engines. Starting from the initial conditions given by $Da = 0$ or $\chi \to \infty$, a decrease of χ, but such that $\chi > \chi_I$ leads to a minor temperature increase. This increase indicates the occurrence of some chemical reactions that correspond to a regime of slow chemistry. The flow time scale (τ_t) is much shorter than the larger chemical time scales (τ_{ch}) and it prevents autoignition. When the Da achieves the critical value Da_I, the mixture ignites. With reference to Diesel combustion, this move along the lower branch of the curve and up to the ignition point corresponds to the interdiffusion of the fuel from the spray with the surrounding hot air that decreases mixture fraction gradient and hence, also the scalar dissipation rate. The ignition of the mixture corresponds to a jump to the upper branch of the S-shaped curve that represents the burning state.[80]

5.2.3 Mixing in the DI-SRM

In the DI-SRM, mixing occurs between the notional particles that compose the in-cylinder mixture. Each particle represents a one-point and one-time PDF for a set of composition variables. The composition variables (ϕ) correspond to enthalpy, $h(t)$, from which the temperature is determined and species mass fraction, $Y_i(t)$. Thus, $\phi(t) = (Y_1, \dots, Y_{N_S}, h; t)$, where $i = 1, \dots, N_S$ and N_S denotes the number of species in the reaction mechanism. Hence, each particle corresponds to a real fluid particle since in the DI-SRM the individual species are fused within the particles as illustrated in Fig. 5.2.

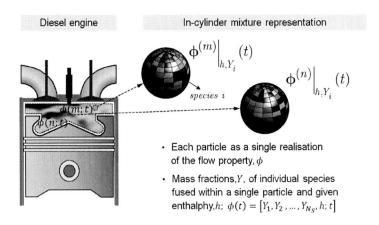

Figure 5.2: Illustration of the in-cylinder mixture in the DI-SRM

In Fig. 5.2, $\phi^{(m)}$ and $\phi^{(n)}$ are two particles that represent two different thermochemical and flow states, respectively before and after the mixing event. Changes in the flow domain result in changes of the species mass fraction and enthalpy of the particles in the gas-phase. Hence, the purpose of mixing modelling is to prescribe the evolution of particles in the gas-phase such that they mimic the composition change of real fluid parcels due to mixing caused by the turbulence. Here, the main task is the determination of mixing time that is needed for the closure of any mixing model. The mixing time decides about frequency of the mixing between the particles and hence, it governs the inhomogeneity of the in-cylinder mixture for species mass fraction and temperature that in some extent represents turbulence-chemistry interaction.

5.3 Representative Mixing Time Model

5.3.1 Modelling Concept and Assumptions

At the microscopic level, mixing corresponds to velocity fluctuations and the occurrence of different time scales in the flow that is expressed by Eq. (5.2). In turn, chemical reactions occur at the molecular level, where mixing time is given by Eq. (5.3). In this work it is assumed that in terms of the history, the mixing time modelled for the DI-SRM should mimic the turbulent mixing time, τ_t, whereas its absolute values should also approximate the effects due to time scales at the molecular level, τ_ϕ. Overall, this is considered as the modelling of microscopic mixing to cover the effects at the molecular level. Furthermore, different values of mixing time should be considered respectively to chemical and physical processes occurring during the engine cycle.

Crank angle dependency of the molecular mixing time during the engine cycle is indicated by the inverse scalar dissipation rate, which is a key quantity in the modelling of non-premixed combustion using the flamelet concept [12, 93, 118]. The inverse scalar dissipation rate represents the time scale of scalar fluctuations [118] and therefore is an indicator of molecular mixing time. In 3D CFD simulations of Diesel engines, employing representative interactive flamelet models (see, e.g., [11, 121]), the scalar dissipation rate is modelled based on the turbulent time scale obtained from the mean turbulent kinetic energy and its dissipation (see Eq. 5.2). By analogy to this approach, a crank angle dependent turbulent mixing time can be used for the closure of mixing models in the DI-SRM of Diesel engine in-cylinder processes.

Using Eq. (5.2) and Eq. (5.3), the Favre averaged scalar dissipation rate can be modelled by an algebraic expression as in [118].

$$\widetilde{\chi} = C_\phi \frac{\widetilde{\epsilon}}{\widetilde{k}} \widetilde{\psi''^2} \tag{5.5}$$

Equations (5.2),(5.3) and (5.5) indicate the relation between the turbulent mixing time of velocity fluctuations (τ_t) and scalar fluctuations (τ_ϕ) that expresses as

$$\tau_\phi = C_\phi^{-1} \frac{\widetilde{k}}{\widetilde{\epsilon}} = C_\phi^{-1} \tau_t. \tag{5.6}$$

In these equations, C_ϕ is a constant of proportionality – *mixing time constant* that is interpreted as the velocity-to-composition decay time. In turn, τ_t/C_ϕ is understood as the intensity of scalar mixing caused by the velocity fluctuations [126]. C_ϕ is not an universal constant [106, 126, 170]. Among others, it depends on the character of scalar fluctuations and the initial flow conditions. Its value may also be affected by the applied mixing model [24] and by the accuracy of τ_t calculations. Previously, and depending on the investigated flow configuration, the values between 0.6 [126] and 4.0 [106] have been used instead of standard value 2.0 [106]. In Large Eddy Simulations (LES), values of up to 10 have been reported [59].

Equation (5.6) is the basic equation for the modelling of mixing time for the DI-SRM. Applying this equation, it is assumed that the turbulent mixing time history of scalar fluctuations, τ_ϕ, which is needed for the DI-SRM, follows, with a certain constant, C_ϕ, the turbulent mixing time of velocity fluctuations, τ_t. The history of τ_t can be obtained from 3D CFD calculations as an averaged history for the entire combustion chamber volume (see Eq. 5.2). By optimising the constant C_ϕ the absolute values of mixing time τ_ϕ should approximate the mixing time effects at the molecular level.

A general trend of the mixing time τ_t is known from existing 3D CFD calculations (see, e.g., Fig. 4.1). Thus, as an alternative to the above approach, the history of mixing time for the DI-SRM can also be obtained via a direct modelling or parametrisation by assuming that in some extent it should follow the trend of τ_t from 3D CFD. These two modelling concepts are outlined in subsequent two subsections. In the context of 0D simulations using the DI-SRM, the modelled mixing time is denoted as *volume-averaged representative mixing time*.

5.3.2 Mixing Time Based on 3D CFD Data

Equation (5.6) states that if the mixing time τ_t is known before simulations, then by calibrating the constant C_ϕ the mixing time for the DI-SRM can be derived. To verify this assumption, exemplary calculations were carried out for Case 1 outlined in Section 5.4. For this case the turbulent mixing time, τ_t, was available from 3D CFD calculations. A close match between the histories of measured and calculated by the 3D CFD model the in-cylinder pressure motivates the use of τ_t as basis for the derivation of the mixing time for the DI-SRM. Through the iterative changes of C_ϕ, the mixing time was calibrated until the best possible match between the experimental and the simulated in-cylinder pressure histories was obtained. This approach can be understood as PDF modelling applied in a post-process. Based on the preliminary tests, C_ϕ was varied between 0.1 and 0.2 with 0.005 interval. The best results were obtained using C_ϕ=0.175 as it is presented in Fig. 5.3 and Fig. 5.4.

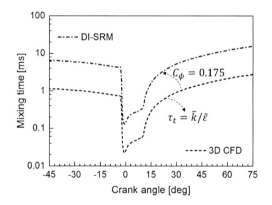

Figure 5.3: Determination of the mixing time based on C_ϕ calibration and given turbulent mixing time, τ_t, that was obtained from 3D CFD computations

Initially, the predicted by the DI-SRM in-cylinder pressure history (see Fig. 5.4) underestimates slightly the histories from measurements and 3D CFD calculations. After approximately 10 °CA ATDC, it traverses these two histories and reaches higher maximum peak value. Subsequently, during the expansion phase, it decreases much faster than the corresponding experimental and 3D CFD counterparts. This fast decrease results from lower RoHR values that in turn is attributed to too slow mixing intensity in that phase (Fig. 5.3). Corresponding deviations are observed for the RoHR histories. The two peaks in the RoHR history in the regime of transition from the premixed to the mixing controlled combustion are not well matched. The RoHR predicted by the DI-SRM has only one peak that over-predicts significantly the reference RoHR trace.

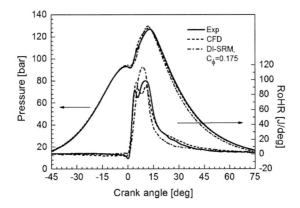

Figure 5.4: Measured and 3D CFD-based in-cylinder pressure and rate of heat release compared to DI-SRM results obtained with C_ϕ=0.175 (see Fig. 5.3)

The inaccuracy of the DI-SRM results (Fig. 5.4) indicates that the determination of the mixing time by optimising only the constant C_ϕ and using an integral volume-averaged turbulent mixing time, τ_t, from CFD requires further consideration. Here, possible reasons for the observed inaccuracy can be due to simplifications made while transforming 3D physical process into 0D modelling framework via τ_t. The mixing time τ_t is not conditional with respect to the mixing process it governs. Furthermore, the shortcomings in the realisation of the mixing process itself (see Section 3.5.5) may also affects the results. Similarly, negative impact can be due to the missing modelling of near-wall processes such as fuel deposition on the wall or flows into and out of crevices. The effect of these processes on mixture composition and specifically HC emission is partially achieved by the calibration of the mixing time history. Hence, mixing time is also a global model parameter for the DI-SRM. Besides governing mixing intensity, it must compensate in some extent the missing effects of 3D geometry on engine in-cylinder processes. In this context, to improve the accuracy of the DI-SRM results, it is necessary to also calibrate/parametrise the shape of the mixing time history as it is presented in next subsection.

5.3.3 Parametrised Mixing Time

The determination of mixing time based on Eq. 5.6 is straight only if the turbulent kinetic energy and its dissipation are known from CFD. Furthermore, the necessity to rely on prior CFD calculations reduces significantly the value of the DI-SRM results. However, an overall trend of τ_t can be deduced from the existing CFD results such as those in Fig. 4.1. Therefore, describing the changes of τ_t by mathematical functions and optimising them, one can directly model the history of τ_ϕ (Eq. 5.3). This corresponds to the optimisation of the constant C_ϕ in Eq. 5.6 having known and unchanged τ_t from CFD.

Following the above reasoning, a parametrised mixing time model has been devised for the description of the mixing process in Diesel engines. The modelling presented in this section[1] is concerned with engines operated with single fuel injection. The mixing time proposed, on a crank angle basis, distinguishes between four regimes of the engine cycle that differ in time scales of the mixing process. It is also considered as a volume-averaged and representative mixing time for the whole combustion chamber. To each regime a separate function is assigned, i.e. τ_0, τ_1, τ_2 and τ_3 as schematically presented in Fig. 5.5. Later on in this work, the functions refer directly to mixing time in these four regimes. The sum of the mixing times τ_0, τ_1, τ_2 and τ_3 defines the joint mixing time history that is used by the DI-SRM. After calibration, the mixing time is denoted by τ. The difference between the current modelling and the one in Section 5.3.2 is that here the mixing time for the DI-SRM can be determined interactively, with no prior knowledge required about the τ_t from CFD.

The regime τ_0 in Fig. 5.5, between the inlet valve closure (IVC) and the crank angle representing the signal that initiates fuel injection (IS), corresponds to the part of the cycle before any fuel has been injected. Here, the mixing intensity is controlled mainly by the internal flow due to the compression process. In this phase, the mixing time is modelled as a constant value linear function τ_0.

In the regime τ_1 fuel is injected under high pressure into the combustion chamber. It vaporises and subsequently mixes with the rest of the in-cylinder gases that is followed by air entrainment into the spray. The mixing time

---

[1]Sections 5.3.3 and 5.4.4 of this chapter are based on an article [112] published in the Proceedings of the 23rd International Colloquium on the Dynamics of Explosions and Reactive Systems (ICDERS), USA, Irvine, CA, July 24-29, 2011.

is represented by a linear function τ_1 that is controlled by τ_{11} and τ_{12}. The duration of τ_1 is governed by the SOI and the EOI. The transition of the mixing intensity between τ_0 at the IS and τ_{11} at the CA of SOI is also modelled by a linear function. The difference between the IS and the SOI is denoted as the injection delay constant.

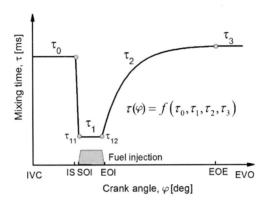

Figure 5.5: Concept of the representative volume-averaged mixing time for DI-SRM-based simulations of Diesel engines operated with single fuel injection

In the regime τ_2 the majority of the chemical energy is released and exhaust emissions are formed. The regime corresponds mainly to the mixing controlled combustion, though the fuel injection, atomisation and vaporisation may still be in progress. Here, the turbulent kinetic energy, which has increased previously due to fuel injection, is dissipated and the mixing time is modelled by an exponential function (see, e.g., [43]). The exponential increase of mixing time begins at the crank angle corresponding to the EOI. The end of the exponential decay of mixing (EOE) is a calibrated parameter. It should be defined so that the τ_2 covers at least the duration of combustion.

Here, it must be remembered that assigning τ_1 with the fuel injection process is considered as an initial state. Following the description in Section 4.3, the fuel vaporisation rate is controlled by the calibration of fuel injection rate profile and mixing intensity. Hence, after the calibration, SOI $\equiv$ SOV and EOI $\equiv$ EOV (see Fig. 4.2 and Fig. 5.5). The modelled SOV determines the anchorage of both the mixing time τ_1 and the vaporisation rate. Furthermore, depending on the thermodynamic conditions in the combustion chamber, the angular position of EOV may also be calibrated (see the discussion in Section 4.3).

As a consequence of the assumptions made, from now on in this work, the parameters SOV and EOV are used instead of SOI and EOI when discussing the timing of the calibrated mixing time.

The regime τ_3 corresponds to the late phase of the engine cycle. Here, the combustion process and pollutants formation have in large part finished. Hence, the modelled mixing time is of less importance. For this reason it is described by a linear function of constant value τ_3.

Following the modelling concept in Fig. 5.5 and remembering about the modelling of fuel injection and vaporisation, the overall mixing time history, τ, for the DI-SRM is obtained by summing up τ_0, τ_1, τ_2 and τ_3. The resulting profile is devised for the description of mixing intensity in Diesel engines operated with single fuel injection that is expressed by Eq. (5.7).

$$
\tau(\varphi) = \begin{cases}
\tau_0, & IVC \leq \varphi < IS \\
\dfrac{\tau_{11} - \tau_0}{SOV - IS}\left(\varphi - IS\right) + \tau_0, & IS \leq \varphi < SOV \\
\dfrac{\tau_{12} - \tau_{11}}{EOV - SOV}\left(\varphi - SOV\right) + \tau_{11}, & SOV \leq \varphi < EOV \\
\tau_3 - \tau_3\exp\left(-B\left(\varphi - EOV\right)\right) + \tau_{12}, & EOV \leq \varphi < EOE \\
\tau_3, & EOE \leq \varphi \leq EVO
\end{cases} \tag{5.7}
$$

In Eq. (5.7), φ represents the angular position during the cycle. The parameter B is given as $B = ln(2)/t_{1/2}$, where $t_{1/2}$ denotes the time needed for the mixing time to reach one half of its final value at the EOE. The value of $t_{1/2}$ depends on τ_{12}, τ_3 and the duration of the exponential part of the mixing time, $\Delta\tau_2$, that is a difference between the EOE and the EOI. The value of $t_{1/2}$ is calculated as $t_{1/2} = ln(2)\Delta\tau_2/ln(\tau_3/\tau_{12})$.

For engines operated with single fuel injection, the solution of Eq. (5.7) requires determining eight constants, namely, τ_0, IS, τ_{11}, SOV, τ_{12}, EOV, EOE and τ_3. The number of the calibrated constants increases further if a multiple fuel injection strategy is applied as presented in Section 5.5. In some circumstances it is possible to reduce the number of the constants down to three for a single fuel injection that is discussed in next section. Furthermore, the constants can be determined in an automated manner using a genetic algorithm. The method of determining the constants of the mixing time model and the computational cost of computations are discussed separately in Section 6.3. In turn, Section 5.4 and Section 5.5 focus on the overall performance of the parametrised mixing time model.

5.4 Application to Engines with Single Fuel Injection

Validation of the parametrised mixing time model has been carried out based on the experimental and 3D CFD data extracted from [104]. The data refer to a Diesel fuelled engine that was operated at speed 2000 rpm, ϕ=0.75, EGR=27% and with single fuel injection (Case 1 in Table B.1 of Appendix B).

Performances of the developed mixing time model were further verified based on the experimental data from Engine B (see Table A.1 of Appendix A). This is an n-heptane fuelled Diesel engine for which the measured exhaust emissions were available. The engine was operated at 2000 rpm, ϕ=0.55, EGR=33% and with single fuel injection (Case 2 in Table B.1 of Appendix B).

For both Case 1 and Case 2, simulations were performed over the closed part of the engine cycle, between IVC and EVO. The setup of the DI-SRM parameters follows the configuration A from Table 4.2. Fuel oxidation and emission formation are modelled using the 121-species n-heptane mechanism taken from [173]. Its exemplary performance is discussed in Section 4.5.

5.4.1 Methodology

The modelling is based on Eq. (5.7) and is applied to Case 1. For engines operated with single fuel injection, the solution of that equation depends on eight constants. This is far too many for an efficient parametrisation of the mixing time. To eliminate this drawback, a best practise has been compiled on how to correlate some of the mixing time constants with each other or assume them before simulations. As a result, the number of constants subjected to calibration has significantly decreased.

Mixing Time τ_0 Between IVC and IS, the mixing time τ_0 depends mainly on the in-cylinder flow that is governed by the compression process. In comparison to other phases of the cycle, such as fuel injection or combustion, here the intensity of mixing is rather low. Its impact on the engine performance is also small since fuel has not been injected yet and mixing occurs only between air and rest gasses, if any. Hence, frequently τ_0 can be assumed before simulations and kept constant. However, if no prior knowledge about the combustion process is available for a given engine operating point, the sensitivity of τ_0 should be verified. The duration of the regime τ_0 is determined by the angular positions of IS and IVC.

Vaporisation Timing Following the modelling assumptions presented in Section 4.3, the angular position of SOV must be calibrated. The position of EOV can be set before simulations based on the assumption that the duration of fuel vaporisation equals the duration of fuel injection, Δ_{inj}, i.e. EOV=SOV+Δ_{inj}. However, if that assumption is not valid, then the EOV must also be adjusted accordingly. Furthermore, a linear transition can be assumed for a mixing time between τ_0 to τ_{11} and equal to 1.0 °CA. This transition represents a delay, in crank angle degrees, between the signal initiating the fuel injection process and the moment at which fuel enters the combustion chamber. Varying the delay between 1 °CA and 3 °CA has not affected simulations. Hence, in this work the delay equal 1.0 °CA has been used as a constant.

Mixing Time Parameters τ_{11} and τ_{12} Mixing time τ_1 between SOV and EOV is governed by τ_{11} and τ_{12} (see Fig. 5.5). In general, the intensity of mixing in this regime may change as the fuel injection and vaporisation processes proceed. If so, then it is assumed that the mixing time changes linearly between τ_{11} to τ_{12} and both of these parameters are to be calibrated. However, if the injected fuel vaporises rapidly, then the assumption that $\tau_{11}{=}\tau_{12} \equiv \tau_1$ may be valid and implies crank angle independence of mixing time in this phase.

Mixing Times τ_2 and τ_3 Mixing time τ_2 is described by an exponential function. The shape of the function is determined by τ_{12}, $t_{1/2}$, τ_3 and the duration of the exponential part of the mixing time, $\Delta\tau_2$ (see also Section 5.3.3). The value of $\Delta\tau_2$ depends on the position of EOE that can be set before simulations by assuming that it occurs shortly after the EOC. In turn, the EOC can be determined from the RoHR history. Since τ_1 is always calibrated, therefore, the calibration of τ_2 depends on $t_{1/2}$ and τ_3. There are two options: either $t_{1/2}$ is calibrated and τ_3 is a result or τ_3 is calibrated and $t_{1/2}$ is a result. With respect to the mixture formation, the most sensible is a regime corresponding to mixing time τ_1 and the part of τ_2 up to $t_{1/2}$, approximately. Hence, the calibration of $t_{1/2}$ is preferred over τ_3.

Depending on engine operating conditions under consideration, the above reasoning enables significant reduction of the number of calibrated parameters of the mixing time model. In an extreme situation and for engines operated with single fuel injection, only three constants have to be calibrated, namely, SOV, τ_1 and $t_{1/2}$ as it is presented in Section 5.4.4.

5.4.2 Validation of the Method

Validation of the parametrised mixing time was carried out against the experimental and the 3D CFD data from Case 1. Case 1 refers to an engine operated with single fuel injection hence, the calibration of mixing time followed Eq. (5.7). Initially, τ_0 was also included in the calibration process. However, several test calculations showed that usually, τ_0 was found between approximately 0.7 ms and 1.1 ms. Therefore, a mean value from that regime (0.9 ms) has been adopted as a known constant, which is very close to the value from the CFD calculation for that operating point (Fig. 5.7). Finally, using the measured in-cylinder pressure history as a simulation target, the mixing time was determined by the calibration of four constants in Eq. (5.7), namely, SOV, τ_{11}, τ_{12} and $t_{1/2}$. The remaining parameters in that equation were set before simulations, based on the methodology from Section 5.4.1. The resulting engine in-cylinder pressure and rate of heat release histories from the DI-SRM are presented in Fig. 5.6. In turn, the plausibility of the calibrated mixing time is discussed separately in next subsection.

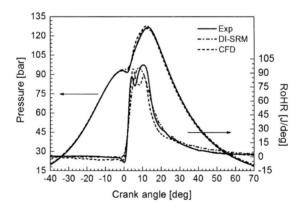

Figure 5.6: In-cylinder pressure and rate of heat release traces from the DI-SRM compared to the experimental data and 3D CFD calculations

Overall, the rate of heat release and in-cylinder pressure histories match reasonably well with the experimental and the 3D CFD data. Similarly as for the results in Fig. 5.4, also here some inefficiency is observed in capturing the two peaks of the RoHR history that correspond to the transition from the premixed to the mixing controlled combustion. These two combustion modes are characterised by different time scales and the simulated mixing time history is somehow a mean representation of the mixing time in that regime. Here however, the simulated maximum peak of the RoHR is much closer to the reference experimental and CFD values. The improvement is attributed to higher accuracy of the prediction of mixing time through its parametrisation that is discussed in next section. Furthermore, the deflection in the experimental RoHR history due to fuel vaporisation, around 0 °CA is not captured by the model. This is contrary to other operating points investigated in this work (see, e.g., Fig. 4.4 and Fig. 6.14).

5.4.3 Mixing Time Constant

By comparing τ_t from CFD with τ_ϕ optimised for the DI-SRM (Fig. 5.7), the reverse engineering process can be verified. The legitimacy of the comparison between mixing times from CFD and DI-SRM gives a close match between the in-cylinder pressure and RoHR histories obtained from both models (Fig. 5.6).

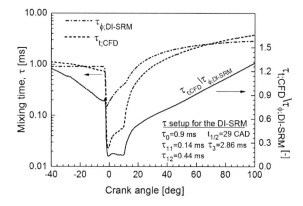

Figure 5.7: Mixing time history from 3D CFD calculations compared to the parametrised history for the DI-SRM (Eq. 5.7) and the ratio between them

The mixing time history from the DI-SRM matches the CFD history qualitatively (results on a logarithmic scale). The difference between them can be accessed through the analysis of the mixing time constant C_ϕ (Eq. 5.6) as it is presented in Section 5.3.2.

The ratio between τ_t and τ_ϕ (Fig. 5.7) varies between 0.14 and 1.30. The smallest values are obtained during the fuel injection and vaporisation, between -2 °CA and 10 °CA ATDC. The highest values occur during early compression and late expansion. The changes of the C_ϕ constant correspond approximately to the values reported in the literature mentioned in Section 5.3.1 ($C_\phi[0.6,10]$). Only the smallest value 0.14 is outside of the reported range. However, it should be noted that the reported C_ϕ values refer mainly to the simulations of simple laboratory flames or non-reactive flows. In such configurations the flow conditions may differ significantly from these found in Diesel engines. Thus, it is concluded that the parametrised mixing time history used by the DI-SRM is plausible.

In Fig. 5.8 the mixing time history obtained by optimising the constant C_ϕ in Eq. (5.6) (see also Fig. 5.3) are compared to the parametrised mixing time history that was calibrated according to Eq. (5.7). During the fuel injection and vaporisation phases (τ_1 in Fig. 5.5), the parametrised mixing time reaches only slightly higher values. This is on the contrary to the remaining parts of the cycle, where substantial differences are observed between the two traces.

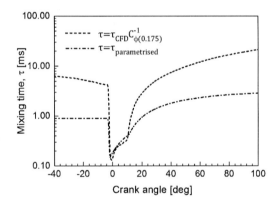

Figure 5.8: Mixing time history for the DI-SRM based on CFD calculations and C_ϕ optimisation in Eq. (5.6) (dashed line) versus the parametrised mixing time history obtained from Eq. 5.7 (dash-dotted line)

Before -2 °CA ATDC, the difference between the two mixing time histories in Fig. 5.8 has no impact on the results since the combustion process has not started yet. After 10 °CA ATDC, the parametrised mixing time reaches significantly lower values. Slightly longer time scales during the fuel injection and vaporisation along with significantly shorter time scales after 10 °CA ATDC, which were obtained from the parametrised mixing time model, yielded an improved prediction of the in-cylinder pressure and RoHR histories when compared to the results in Fig. 5.4. Specifically, the prediction of the maximum peak of RoHR history has been improved. Hence overall, it is concluded that the 3D effects of the mixing process are accurately represented within 0D domain by the parametrised mixing time model.

5.4.4 Model Performance

To further analyse the parametrised mixing time model, its performance has been compared to the results obtained with constant value mixing time. The results presented refer to Case 2 from Table B.1 of Appendix B.

Crank Angle Dependency of the Mixing Time

By using a constant value mixing time it is assumed that the mixing processes in the combustion chamber are characterised by the same time scale regardless of the angular position during the cycle. This concept originates from the application of the SRM to HCCI engines (see Section 1.2.2). With respect to Diesel engines, such a modelling introduces a simplification that implies the mixing time be somehow a mean or characteristic mixing time that approximates different time scales occurring in the Diesel cycle. To verify this modelling concept, the single value mixing time was varied between 0.04 ms and 10 ms. The best match for the in-cylinder pressure, rate of heat release and exhaust emissions was obtained for $\tau=0.3$ ms. In Fig. 5.9 and Fig. 5.10, the results obtained are compared to the experimental data and the results from the crank angle dependent mixing time model.

Until approximately 6 °CA ATDC the simulated in-cylinder pressure and rate of heat release accurately match the experimental data for both mixing time histories. Then, the results from the constant value mixing time indicate slower combustion, which results in lower pressure and RoHR gradients. The maximum peak values of pressure and RoHR are moved to the later part of the cycle. This is contrary to the results obtained with the crank angle dependent mixing time. The start of combustion, peak in-cylinder pressure and its angular position in the cycle match more accurately the experimental counterparts. Slight over-prediction is seen after the maximum pressure peak towards EVO.

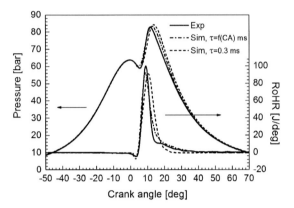

Figure 5.9: Simulated and experimental histories of in-cylinder pressure and RoHR based on the constant value and crank angle dependent mixing time

The difficulty to match the simulated results with the experimental data through the calibration of the constant value mixing time is more enhanced for engine exhaust emissions shown in Fig. 5.10. The under-prediction of exhaust NO_x for $\tau=0.3$ ms is attributed to the lower maximum peak of the rate of heat release. In turn, higher values of the rate of heat release and in-cylinder pressure during the later part of the cycle (after approximately 10 °CA ATDC in Fig. 5.9) may indicate more complete fuel oxidation. As a consequence, the unburned HC occur in lower concentration at EVO.

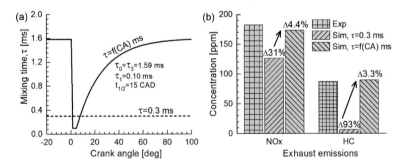

Figure 5.10: Constant value and crank angle dependent mixing time histories (a) and corresponding engine exhaust NO_x and HC (b) that are plotted over the measured data; by Δ is denoted the relative difference between the simulated and the experimental data

Even if one assumes the predicted in-cylinder pressure and rate of heat release, using τ=0.3 ms, as acceptable for 0D simulations, the corresponding NO_x and in particular exhaust HC are predicted with too low accuracy. This indicates that the constant value mixing time may not adequately mimic the locality of the mixing process in Diesel engines. On the contrary to that, the crank angle dependent mixing time improves significantly the accuracy of the predicted exhaust NO_x and HC. The relative difference between the simulated and the experimental data is 4.4% and 3.3% for NO_x and HC, respectively.

Local In-Cylinder Conditions

Further insight into the difference between the constant value and crank angle dependent mixing time is gained from the analysis of the local in-cylinder temperatures and equivalence ratios. These properties determine the state of the mixture that decides about the local rates of heat release and pollutants formation. The change of temperature and composition for each particle results from mixing, heat transfer and fuel injection. The combined influence of these processes on particle properties is tracked on the ϕ-T diagrams in Fig. 5.11. Exemplary results are presented for 7 °CA and 25 °CA ATDC. At 7 °CA the results from both method start to deviate. The map at 25 °CA represents the local conditions towards the end of combustion. The mixture corresponding to a constant value and crank angle dependent mixing time are denoted respectively as mixture A and mixture B.

At the selected angular positions mixture A is more homogeneous than mixture B in the sense that the corresponding PDFs reach higher maximum peaks and the distributions are narrower. At 7 °CA the maximum peak value of ϕ is located before ϕ=0.55 that is an overall equivalence-ratio of the in-cylinder mixture. This indicates mixture over-leaning due to rather intensive mixing at earlier times during the cycle. On the ϕ-T map, this is represented by a cloud of particles around 1000 K. It is expected that more lean particles with lower temperature is a reason for a lower concentration of NO_x for mixture A (Fig. 5.10b, τ=0.3 ms).

In the later part of the cycle, at 25 °CA, the PDF distributions are more similar to each other. The mixture A is still more premixed, but the maximum peak values are located for both mixtures at ϕ being close to the global equivalence-ratio (ϕ=0.55). At 25 °CA ATDC the combustion approaches an end but still the oxidation of the remaining hydrocarbons is possible. This, together with less spread distribution of mixture A, yields a lower concentration of HC at EVO (Fig. 5.10b).

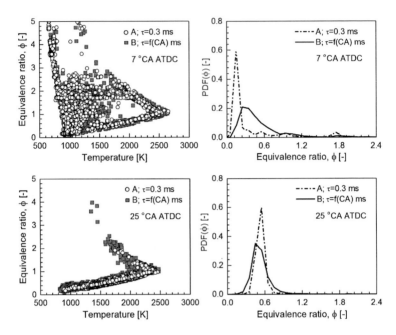

Figure 5.11: Exemplary ϕ-T maps and PDF distributions of ϕ from the DI-SRM at different instants during the cycle. Top: 7 °CA ATDC. Bottom: 25 °CA ATDC; circles/dash-doted lines refer to the results from the constant value mixing time and squares/solid lines from the crank angle dependent mixing time

Parameter Studies

A model parameter studies were performed to further analyse the impact of the modelled mixing time on the in-cylinder performance parameters. Specifically, the influence of τ_1 and $t_{1/2}$ (see Eq. 5.7) on the in-cylinder pressure, combustion progress, exhaust HC and NO_x was investigated. These parameters are most sensitive to the mixing time changes.

Mixing time history from Fig. 5.10, which corresponds to the engine results in Fig. 5.9, was taken as a baseline for the analysis. The history is defined by τ_1=0.1 ms, $t_{1/2}$=15 °CA ATDC and τ_0=τ_3=1.59 ms. Calculations were carried out for three different values of τ_1 and $t_{1/2}$. Figure 5.12 and Table 5.1 summarise the results obtained.

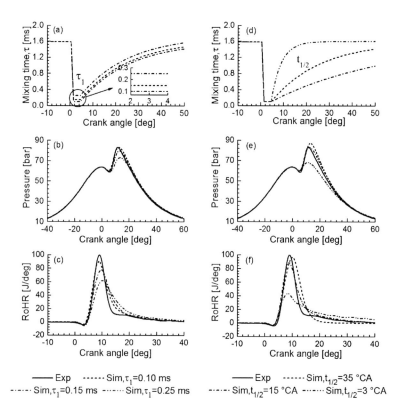

Figure 5.12: Sensitivity study of the impact of mixing time model parameters, τ_1 and $t_{1/2}$, on the simulated histories of in-cylinder pressure (b, e) and rate of heat release (c, f). Left: the impact of τ_1 (a). Right: the impact of $t_{1/2}$ (d)

Mixing time τ_1 appears most influential on the DI-SRM results. Even a very small change (0.05 ms) has an impact on the in-cylinder pressure peak, combustion rate and exhaust emissions. A general tendency is that an increase of τ_1, which also slightly increases $t_{1/2}$, delays the SOC that is defined by the CA_{10}. As a result, the maximum pressure peak decreases and occurs later during the cycle. The combustion duration is longer and the concentration of exhaust HC increases. Similar impact on the in-cylinder pressure and exhaust HC has the advancing of $t_{1/2}$. The change of $t_{1/2}$ has almost no impact on SOC. Furthermore, only very early position of $t_{1/2}$ influences the angular position of the combustion centre that is given by CA_{50}.

Table 5.1: Measured exhaust emissions and combustion progress compared to the DI-SRM simulations from different mixing time setups

Parameter	Exp	τ_1 [ms]			$t_{1/2}$ [°CA]		
		0.10	0.15	0.25	35	15	3
NO_x [ppm]	182	174	188	175	86	174	155
HC [ppm]	87	90	145	210	9	90	691
CA_{10} [°ATDC]	7.2	7.1	7.4	8.0	7.2	7.1	7.2
CA_{50} [°ATDC]	9.9	10.8	11.6	13.2	10.8	10.8	16.0
CA_{90} [°ATDC]	18.6	20.2	23.6	30.2	15.5	20.2	44.6

Overall, decreasing τ_1 and increasing $t_{1/2}$ increases the in-cylinder pressure and decreases the unburned HC. A contrary effect is observed for longer τ_1 and shorter $t_{1/2}$. This correlation cannot be stated for exhaust NO_x (see Table 5.1). Mixing time that retards the location of CA_{50} and decreases the maximum pressure peak does not always reduce the concentration of NO_x. An interaction between the formation of NO and NO_2 that compose NO_x can be a possible explanation of this behaviour. At a certain temperature window, frequently during the expansion stroke, the fraction of NO_2 in NO_x increases at the expense of NO. Since the NO_2 molecule is heavier than NO, the resulting mass-based concentration of NO_x increases. Depending on the local temperatures inside the cylinder this interaction can increase or decrease the total concentration of NO_x. These aspects of NO_x formation are discussed in more detail in Section 7.3.

5.5 Application to Engines with Double Fuel Injection

The mixing time model is further investigated with respect to the simulation of Diesel engines operated with double fuel injection. A strategy is considered with a short time-gap between the pilot and the main injection. The pilot injection is significantly smaller than the main injection. Such strategies are frequently employed for noise reduction and the pilot injection is used for preconditioning the combustion chamber for main injection and combustion.[1]

---

[1]This section is based on an article [110] published in the Proceedings of the 1st International Conference on Engine Processes, ISBN 78-3-8169-3222-2, Berlin, Germany, June 6-7, 2013.

5.5.1 Overall Assumptions

The modelling follows the ideas developed for the simulation of single fuel injection strategy. The parametrised mixing time, which is governed by Eq. (5.7), is applied to both injections. The resulting mixing time history is shown schematically in Fig. 5.13. The two minima (τ_1^A, τ_1^B) of the profile correspond to the pilot and the main fuel injection, respectively. They perform in a similar manner as for the single fuel injection that is discussed in previous subsections. Consequently, τ_1^A and τ_1^B are governed by τ_{11} and τ_{12} for both pilot and main injections (see also Fig. 5.5 and Eq. 5.7).

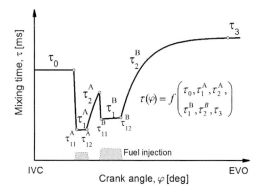

Figure 5.13: Concept of the representative volume-averaged mixing time for DI-SRM simulations of Diesel engines operated with double fuel injection

The scaling of the mixing time between the two minima accounts for the differences in the injection duration and rate. The model can be applied to operating modes where the pilot and main injection occur with substantial delay. By further reproducing of the mixing time profiles, it is possible to cover multiple fuel injection.

5.5.2 Simulation Results

Exemplary test calculations were carried out for Engine C specified in Table A.1 of Appendix A. It is a Diesel fuelled engine that was operated with double fuel injection at low speed and medium load (Case 3 in Table B.1 of Appendix B).

The setup of the DI-SRM follows configuration C from Table 4.2. The 121-species n-heptane mechanism from [173] is used as a Diesel surrogate. To account for the differences in lower heating value between the Diesel fuel used during the experiment and n-heptane, the mass of fuel injected during simulations was corrected to ensure the same amount of energy release as in the experimental setup.

To reduce the number of calibrated constants for the mixing time model, the methodology presented in Section 5.4.1 was applied to both pilot and main injection. As a result, the mixing time was determined by calibrating five model constants, namely, τ_1^A, $t_{1/2}^A$, τ_{11}^B, τ_{12}^B and $t_{1/2}^B$ that are shown in Fig. 5.14. The value of τ_0 was estimated during the preliminary study and τ_3 was calculated from τ_{12}^B and $t_{1/2}^B$.

Simulations focused on verifying whether the mixing time history modelled according to Fig. 5.13 can capture the rate of heat release and in-cylinder pressure governed by the double fuel injection. The results obtained are presented in Fig. 5.14 and Fig. 5.15.

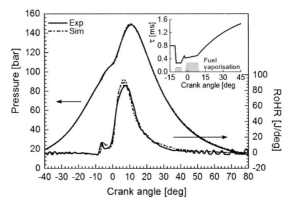

Figure 5.14: Modelled mixing time history (upper right plot) and a comparison between the computed and the measured pressure and rate of heat release traces

Overall, the simulated histories of in-cylinder pressure and RoHR match the experimental counterparts with high accuracy. The first peak in the RoHR history due to pilot injection, between -9 °CA and -1 °CA ATDC, is well captured by the model, proofing thus its plausibility.

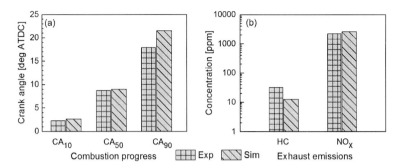

Figure 5.15: Comparison between the computed and the measured data for combustion progress (a) and exhaust emissions (b)

In comparison to the results obtained in Section 5.4 for Case 2, here slightly lower accuracy has been obtained for exhaust emissions. The relative error between the measured and the simulated data is 14% for NO_x and 61% for unburned HC. The over-prediction of exhaust NO_x is attributed to a slightly higher maximum peak of the simulated RoHR. For the same reason, the simulated concentration of unburned HC at EVO under-predicts the experimental data. Lower concentration of exhaust HC is also a consequence of slightly longer combustion duration (Fig. 5.15a) that may promote more complete oxidation of HCs. Further contribution can be due to the differences between the actual Diesel fuel used during the experimental works and n-heptane used as surrogate during simulations. Finally, it should also be noted that the targeted exhaust HC is just 32 ppm and hence, already a few ppm difference increases significantly the relative error between the simulated and the measured data.

5.5.3 Practical Aspects of Simulations

By duplicating the baseline mixing time profile to capture the effects due to multiple fuel injection, the number of calibrated constants of the mixing time model is increased. Consequently, even if one applies simplified modelling from Section 5.4.1 and calculations are carried out in an automated manner using the method presented in Section 6.3, the overall calibration process may still be complex and computationally expensive. This is the main difficulty of the presented method that becomes more noticeable if the number of injections increases above two. However, a model parameter studies have revealed that this drawback can frequently be avoided in practical applications.

Simplifications are possible for engines operated with (a) very short pilot injection, (b) very short post injection and (c) if the pilot and main injection events take place in rapid succession. If the duration of the pilot injection is much shorter that the main injection and additionally the pilot injection involves a small fraction of the total mass of the fuel injected, then at sufficiently high temperature fuel vaporisation and subsequent mixing are fast. The short time needed for these processes makes the exponential decay of the mixing time be not sensitive for the combustion process. In consequence, mixing intensity during the pilot injection can be obtained by calibrating only the mixing time τ_0 (see Fig. 5.5 and Fig. 5.13). Similar reasoning is applied for the operation with short post injection, except that $t_{1/2}$ must be calibrated. If the pilot and main injection events occur in rapid succession, then by a suitable adjustment of the slope of τ_1 (or τ_{11} and τ_{12}) it is still possible to account for different mixing time scales during the pilot and main injection. A practical consequence of such simplifications is the possibility to approximate mixing time scale for multiple fuel injection in a similar manner as for single fuel injection. The overall simulation process is less complex and requires less CPU time.

To demonstrate the performance of such a simplified modelling, exemplary calculations were carried for Case 3 described in Appendix B. This is the same engine case as the one used in the previous section. Also, the configuration of the DI-SRM remains the same as in Section 5.5.2. Considered is an engine operated with double fuel injection and with pilot and main injection events taking place in rapid succession. Therefore, it is assumed that the mixing time can be approximated by only one linear function, τ_1, that covers both the pilot and the main injection (see Fig. 5.5 and Fig. 5.13). The duration of τ_1 corresponds to the total duration of the injection process, including the gap between the pilot and the main injection. The slope of the function τ_1 is expected to represent the mean value of the mixing time profiles that are obtained by two functions τ_1^A and τ_1^B in Fig. 5.13. This simplification has reduced the number of calibrated parameters of the mixing time model to four, namely, SOV, τ_{11}, τ_{12} and $t_{1/2}$ (see Fig. 5.5 and Fig. 5.13).[1] With respect to computational cost (see Section 6.3.4), such a reduction corresponds approximately to twice faster calibration process. The resulting reduced mixing time history and its detailed counterpart are presented in Fig. 5.16. This

---

[1] Henceforth in this section, the mixing time model after simplification is referred to as the reduced model. In turn the full model described in the previous section is referred to as the detailed model.

figure contains also a comparison between the simulated and the experimental histories of in-cylinder pressure and rate of heat release.

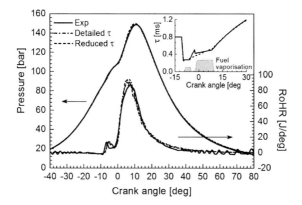

Figure 5.16: Experimental pressure and rate of heat release compared to the simulations based on the detailed (dash-dotted lines) and the reduced modelling of mixing time history (dashed lines)

An application of the reduced mixing time model has led to similar histories of the in-cylinder pressure and RoHR as the ones obtained with the detailed model. As far as exhaust emissions are concerned, the relative error between the measured and the simulated data from the reduced model is 12% for NO_x and 69% for unburned HC. This is also very similar to the values obtained with the detailed model (14% for NO_x and 61% for unburned HC).

Further confirmation of the validity of the reduced modelling of mixing time is gained from the ϕ-T maps and PDFs of ϕ at two instants during the cycle that are presented in Fig. 5.17. At -3 °CA ATDC there is the largest difference between the detailed and the reduced mixing time histories. In turn, 8 °CA ATDC corresponds to the end of the intensive mixing phase τ_1.

Overall, the distribution of particles from both models agree for most of ϕ and T values. Small discrepancies are visible under very rich conditions at 8 °CA ATDC. Fewer particles and relatively smaller discrepancies between them are seen at -3 °CA ATDC. Here, only a small amount of fuel from the pilot injection is still available for the mixing process and most particles are already mixed with air.

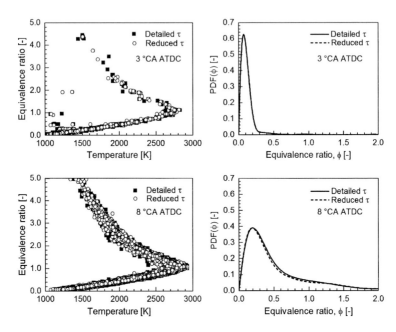

Figure 5.17: Exemplary ϕ-T maps and PDF distributions of ϕ at 3 °CA ATDC (top) and 8 °CA ATDC (bottom); squares/solid lines: detailed τ, circles/dashed lines: reduced τ

It is difficult to quantify the influence of different mixing time histories since the discrepancies between particle-based results from both models are very small. The effects of the stochastic mixing process, heat transfer and fuel injection seem equally contributing to the properties of the particles at these crank angles. This is confirmed by the histories of PDF distributions of the equivalence ratio.

Using as reference the results from the detailed modelling of mixing time, the particle-based analysis lead to the conclusion that the simplification made in the reduced model around the intensive mixing phase τ_1 does not negatively affect the local conditions represented by ϕ and T. Hence, the reduced modelling is considered plausible. On the other hand, it should be remembered that the reduced modelling is not regarded general. It is demonstrated that the method can be applied to model engine with multiple fuel injection. However, its validity is engine and operating point sensitive.

Chapter 6

Engine Simulation Framework for the DI-SRM

6.1 Introduction

The modelling and simulations presented in Chapter 4 and Chapter 5 relate to the improvement of the performance of the DI-SRM by the application of the crank angle dependent volume-averaged representative mixing time model. These studies have not addressed the issue of effective determination of the mixing time history, when it is not available from measurements or 3D CFD calculations; the mixing time must be calibrated for the DI-SRM since 0D models cannot mimic the geometry impact on the mixing process. Another question that remains unanswered is how to transfer the setup of the mixing time from one set of operating points to other ones that differ by speed, load or fuel injection strategy. To these modelling aspects is devoted the present chapter. First, an overall concept of an engineering process employing the DI-SRM for Diesel engine performance simulations is introduced. Then, an automated method of determining mixing time history is proposed. Subsequently, a concept of an engineering process is developed, which allows for predictive simulations of Diesel engine performance parameters at different operating points.

6.2 Simulation Process

An overall process of Diesel engine simulations using the DI-SRM is composed of three main phases as schematically illustrated in Fig 6.1.

Heat Release Analysis (I) An overall engine operating characteristic, including load, EGR and fuel injection strategy is analysed. Based on the in-cylinder pressure data, heat release analysis is performed to quantify the combustion process. Subsequently, the initial conditions, such as pressure, temperature and gas composition at IVC, are determined for engine simulations using the DI-SRM.

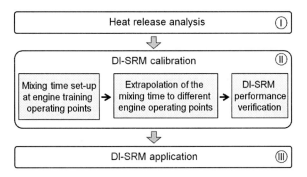

Figure 6.1: Overall engine simulation process using the DI-SRM

DI-SRM Calibration (II) Simulations in this phase are built around the concept of volume-averaged representative mixing time introduced in Chapter 5. Overall, the process is carried out in three steps.

- Determination of mixing time histories at the selected engine operating points using the self-calibration method that is introduced in Section 6.3.1. Histories of the mixing time from different engine operating points indicate overall sensitivity of the modelled mixing time to the changes of engine operating parameters. This is considered as the DI-SRM training.

- Parametrisation of mixing time parameters from the trained model with known engine design variables such as load, speed and fuel injection strategy. The knowledge about the dependencies between mixing time parameters and engine design variables allows the extrapolation of mixing time model setups for the simulation of engine load-speed maps.

- Verification of the parametrised mixing time with respect to the simulation of Diesel engine performance parameters at operating points that were not used during the DI-SRM training.

DI-SRM Application (III) Prediction of engine in-cylinder performance, mapping of engine-out emissions or studies of engine-fuel interactions are among possible applications of the process.

The present chapter is devoted to the second phase of the process shown in Fig 6.1. In Section 6.3 an automated procedure is introduced for mixing time determination. Section 6.4 describes the parametrisation of the mixing time with known engine design parameters.

6.3 Automated Method of Mixing Time Determination

The determination of mixing time histories is understood as an extended heat release analysis. The process relies on the optimisation of parameters governing the mixing time profile until the simulated in-cylinder pressure, rate of heat release and exhaust emissions match the experimental counterparts. The present section demonstrates how the process can be applied to determine the mixing time profile for an engine operated with single fuel injection.[1]

6.3.1 Self-Calibration Procedure

The purpose of the self-calibration procedure is to determine the parameters governing mixing time history such as τ_0, IS, τ_{11}, SOV, τ_{12}, EOV, EOE and τ_3 (see Fig. 5.5 and Eq. 5.7). These parameters are referred to in this section as *design variables* of the DI-SRM. The calibration is targeted at matching the in-cylinder pressure and engine exhaust NO_x and HC from the DI-SRM to the corresponding experimental data.

The in-cylinder pressure values at three representative CA positions are selected as targets during mixing time optimisation. In Fig. 6.2, the targets are denoted by S_1, S_2 and S_3. They are positioned approximately in the regime corresponding to fuel injection and combustion, which are two of the most challenging to accurately capture by simulations.

---

[1]This section is derived in part from the SAE Technical Paper 2012-01-1072, 2012, DOI:10.4271/2012-01-1072 [113].

The first target S_1 is defined slightly after the SOC. Based on the pressure history, the position of S_1 may be roughly estimated as corresponding to the CA where a visible pressure increase is observed that is not due to compression. Alternatively and more accurately, the position of SOC can be determined from the rate of heat release history. Defining the pressure targets before the SOI is not necessary since the compression pressure is usually well predicted if correct information is given about the engine geometry, pressure and temperature at IVC. The target S_2 is defined at the CA corresponding to the maximum peak pressure due to combustion. The target S_3 is defined for the expansion phase, at the CA where it is expected that more than 50% of the energy due to combustion has been released. Matching the pressure at S_1 ensures correct SOC that affects the in-cylinder pressure in the later phase of combustion. Matching the pressure at S_2 ensures correct pressure gradient between the SOC and the maximum peak pressure. In turn, matching the pressure at S_3 prevents increasing the pressure after passing the CA of target S_2. Hence, it ensures the pressure is decreasing during the late combustion and expansion stroke.

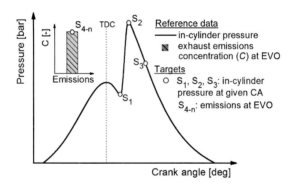

Figure 6.2: Objective function for the determination of mixing time history

The presented method of defining targets S_1, S_2 and S_3 is general. It can be applied to cases with combustion initiated before and after TDC. The exact choice of the positions of S_1, S_2 and S_3 depends on the considered engine case. If necessary, the number of pressure targets can be increased.

Good agreement between the measured and the simulated in-cylinder pressure histories does not necessarily imply a good agreement between the simulated and the measured exhaust emissions (see Section 5.4.4). Pollutant formation is more sensitive to the locality of mixing process than to the mean in-cylinder pressure and temperature. Hence, if simulations are also targeted at predicting exhaust emissions, then it may be necessary to also set their exhaust concentrations as targets during the optimisation of mixing time history. In Fig. 6.2 the emission target is denoted by S_{4-n}, where n denotes the number of target-emissions. Typically, concentrations of HC, NO_x, CO or soot at EVO can be selected as target-emissions. Their choice depends on the overall objective of the calibration process.

The preliminary works have indicated that the unburned HCs are most sensitive to the mixing time history. More recently performed investigations [48] show similar sensitivity of soot. Exhaust NO_x in contrast, when calculated based on the thermal NO mechanism, shows lower sensitivity. Frequently, it can be accurately predicted if the simulated mean in-cylinder pressure and temperature histories match the measured counterparts. Usually, this occurs if properties of the fuel used during experimental work are accurately represented by the surrogate fuel used during simulations.

6.3.2 DI-SRM and Genetic Algorithm

Optimisation of the mixing time history is automated. An integrated tool is applied that combines the DI-SRM and a genetic algorithm (GA). Details of the GA are beyond the scope of this work and can be found in the existing vast literature. In this section, based on [56, 143], only the most important features of GAs are outlined for readability of the section.

Genetic Algorithm

The foundation of any implementation of a GA is the evolutionary theory that describes a process of change or creating of species that follows natural selection. In natural selection the strong species have a greater chance to pass their genes to future generations and in consequence to survive. This process occurs on the expense of weak species that are convicted for extinction.

In the GA nomenclature, an *individual* (*chromosome*) is a primary element that represents a single solution (or design variable) in the solution space. Encoding is used to represent individuals in the solution space. An individual is composed of single elements, called *genes* that control its features. A collection of individuals defines a *population*. Within a population, a *fitness function* is used to numerically access the quality of each individual, i.e. to determine how well a given single solution matches the assumed target. The number of individuals in the population is arbitrarily defined. The fitness function is associated with a *target function* that represents an overall optimisation target. Typically, the GA operation is initiated by generating an initial population and calculating the fitness for each individual. Next, among the individuals in the existing population two are selected that are called *parents* and their *offspring* is created. The selection of parents is random, but is influenced by the values of the fitness function for the individuals; a better fit individual has higher probability to be selected as a parent. Typically, the changes within the population, i.e. selection of parents, generating their offspring and then creating a new population is obtained using operators such as *crossover* and *mutation*. The crossover is a combination of two individuals to form a new individual; by repeating iteratively the crossover, the genes of better fit individuals are expected to appear more frequently in the population. Mutation is an operation of introducing random changes into the offspring created by crossover.

The main idea behind GA operations is an expectation that in subsequent iterations the populations will contain the better fit individuals hence, the fitter overall solution. Usually, GA operations are terminated when a maximum number of iteration has been reached or if in a few subsequent iterations there are small changes in the fit for the best individual.

GA and DI-SRM Coupling

Figure 6.3 shows schematically a concept of coupling between the GA and the DI-SRM. In this configuration, a user-coded function plays a central role. It governs the communication between the GA and the DI-SRM and carries out pre- and post-processing tasks.

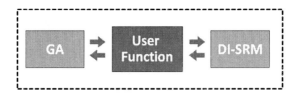

Figure 6.3: Communication between GA and DI-SRM

The structure in Fig. 6.3 is general. Respectively to the functionality of the user function, the integrated DI-SRM and GA model can be applied for solving various tasks related to the simulation of engine in-cylinder processes. In the present work, the user function is tailor-made for the optimisation of a mixing time profile for use in DI-SRM simulations.

Designing the mixing time, such that for the given targets S_{1-n} (see Fig. 6.2) the differences between the simulated and the experimental data reach minima, is a global objective of the integrated DI-SRM and GA. The objective is defined as a minimisation of the function $T(S_i)$ that with reference to Fig. 6.2 is expressed as

$$T(S_i) = \sum_{i=1}^{n} \left(|S_{i,ref} - S_{i,sim}| \right). \tag{6.1}$$

In this equation index i corresponds to the number of targets. The suffices *ref* and *sim* denote reference and simulated data, respectively. In this work, the targets S_{1-n} are equally weighted within the function $T(S_i)$. Equation (6.1) is solved until the simulated engine performance parameters match the selected targets (Fig. 6.2) within a user-defined tolerance. Hence, globally the optimisation of the design variables is considered as a search type problem since there can be more than just one solution fulfilling the assumed tolerance.

6.3.3 Validation of the Method

Exemplary calculations were carried out for Engine B specified in Table A.1. It is an n-heptane fuelled Diesel engine that was operated with single fuel injection at low speed and load (Case 2 in Table B.1 of Appendix B).

The setup of DI-SRM follows the configuration A presented in Table 4.2. The 28-species n-heptane mechanism, described in Section 4.5.1, was applied as Diesel surrogate fuel. Its small size and thus computational efficiency makes it suits the investigations focusing on simulation method development.

The engine operating point considered here is the same as the one used in Section 5.4.4 for which the mixing time profile is fully determined by defining the SOV, τ_1 and $t_{1/2}$.

DI-SRM and GA Setup

Three pressure targets and exhaust HC were selected as targets for matching the simulated data with the corresponding experimental counterparts.

Table 6.1: Optimisation targets according to the concept in Fig. 6.2

No.	Target	Position [°CA ATDC]	Value [bar][1][ppm][2]	Tolerance [%]
S_1	In-cylinder pressure	5	59.0[1]	2
S_2	In-cylinder pressure	12	83.1[1]	2
S_3	In-cylinder pressure	15	76.7[1]	2
S_4	Concentration of HC	EVO	87.0[2]	9

For the considered engine operating point, n-heptane was used during both experimental work and simulations. Based on the preliminary calculations (see also the discussion in Section 6.3.1, p. 103), exhaust NO_x has been excluded from the list of targets. Optimisation of the mixing time was carried out until the targeted in-cylinder pressures (S_{1-3}) matched the experimental values within 2% tolerance. For HC concentration (S_4), a 9% tolerance has been assigned that corresponds to the measurements accuracy. The tolerances are expressed as relative standard deviation.

The mixing time history is determined by optimising τ_1, $t_{1/2}$ and SOV in between the upper and the lower bound (*search space*) that are defined in Table 6.2. The bounds were defined based on the preliminary calculations. In the current configuration of the GA, the design variables are modified within populations containing 10 individuals, i.e. 10 combinations of the mixing time parameters and corresponding to it 10 solutions with respects to the targets S_{1-4} in Table 6.1.

Table 6.2: Search space for the optimised mixing time parameters

No.	Mixing time parameter	Unit	Lower bound	Upper bound
1	τ_1	ms	0.05	0.3
2	$t_{1/2}$	°CA	5	30
3	SOV	°CA ATDC	-1.0	1.0

Calibration Process

The mixing time histories result from the optimisation of the design variables listed in Table 6.2 and with consideration for the targets in Table 6.1. The calibration process is illustrated in Fig. 6.4.

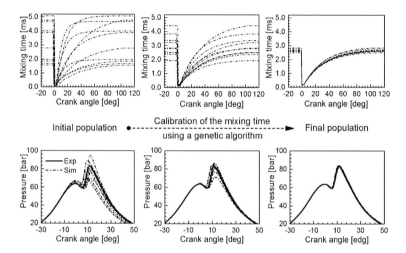

Figure 6.4: The Concept of DI-SRM calibration using a GA. Top: subsequent populations of the mixing time history. Bottom: resulting in-cylinder pressure; solid lines: experiments, dash-dotted lines: simulations

The highest dispersion of the simulated pressure histories is observed for the first populations that represent the initial solution field. As the optimisation of the design variables proceeds, the dispersion between mixing time traces becomes smaller. At the same time, pressure histories approach the target history that is represented by a solid line in Fig. 6.4.

In more detail, the evolution of the optimisation process is illustrated in Fig. 6.5 that shows simultaneously the fitness of all individual solutions of the design variables to all four targets from the Table 6.1. After the 12th population no visible changes in accuracy of the solution was observed. For each target S_{1-4}, single-solutions are represented by filled squares that are connected by a dotted line to visualise the evolution of the optimisation process. The targets are denoted by a dash-dotted line. In each sub-figure, the final result is denoted by an empty square and indicated by an arrow.

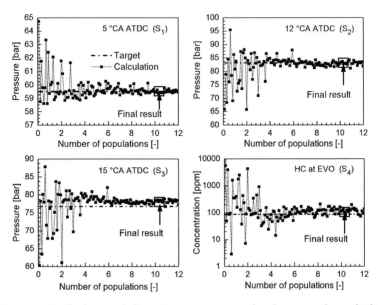

Figure 6.5: Evolution of the optimisation process for the targeted in-cylinder pressures (S_{1-3}) and exhaust HC (S_4) (see Table 6.1) from 12 populations

The evolution of solution shows different sensitivity to the optimisation process for each target. The most stable are solutions for S_1 (the in-cylinder pressure at 5 °CA, Table 6.1). After already the 3rd population, all the results are within the assumed tolerance. For the remaining two pressure targets (S_{2-3}), higher dispersion from the reference data is observed. The unburned HC emission (S_4) is the most sensitive to the optimisation process. The high dispersion from the reference value is especially observed for the first 4 populations. This tendency also confirms a plausibility of choosing the HC as an important target for mixing time calibration.

Fast convergence of the solution is observed for all four targets. After generating approximately the 4th population, the simulated in-cylinder pressures (S_{1-3}) and exhaust HC concentration match reasonably well the assumed targets. Hence, it may not always be necessary to generate many populations to find an acceptable solution. Depending on the GA operators and initially generated population, a solution satisfying the assumed targets may already be found within the initial populations. This is one of the characteristic features of GAs applied for the solution of the search type problems.

In-Cylinder Performance and Exhaust Emissions

An individual 104 found in the 10th population is the first solution that fulfils all targets in Table 6.1. For the selected three pressure targets, the relative difference between the measured and the experimental data is 0.7%, 0.2% and 2% for S_1, S_2 and S_3, respectively. As a consequence, good agreement between the measured and the simulated quantities is observed for the entire cycle (see Fig. 6.6 and Fig. 6.7a).

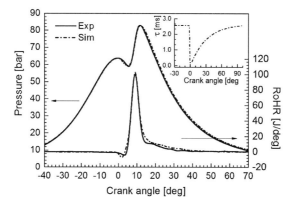

Figure 6.6: Comparison between the computed and the measured in-cylinder pressure and RoHR traces; simulations based on the calibrated mixing time that is described by SOV=-0.26 °CA ATDC, τ_1=0.2 ms and $t_{1/2}$=21.45 °CA

The simulated exhaust HC and NO_x match the measured data with high accuracy (Fig. 6.7b). Relative error between the simulated and the measured data is 4.9% for NO_x and 4.6% for HC. For HC, high accuracy is a result of its consideration as the optimisation target. Furthermore, the same type of fuel was used during the experimental work (n-heptane) and simulations (reaction mechanism for n-heptane), yielding high consistency between the data compared.

Exhaust NO_x was excluded from the target parameters during optimisation. Its prediction results from matching the in-cylinder pressure and exhaust HC. The results obtained indicate that for the 28-species n-heptane mechanism, the formation of NO_x is less sensitive to the mixing time history than HC, for which higher spread of the results is observed in Fig. 6.8 for the same changes of the mixing time history (see Fig. 6.4).

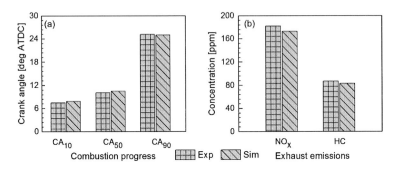

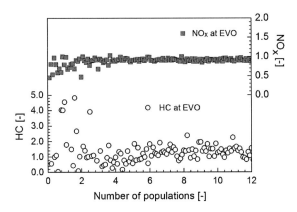

Figure 6.7: Comparison between the computed and the measured data for combustion progress (a) and exhaust emissions (b)

Figure 6.8: Evolution of the exhaust HC and NO_x concentrations from 12 populations; results normalised to the experimental values in Fig. 6.7b

The results in this section are quantitatively similar to those in Section 5.4.4, which were obtained for the same engine operating point, but using the 121-species n-heptane mechanism. By optimising the mixing time, possible negative impact of less accurate reaction mechanisms (the 28-species mechanism, see also Section 4.5.2) can be, into some extent, hidden within the mixing time history. The optimiser searches for a solution without constrains imposed on the mixing time. This further indicates that mixing time, besides governing mixing intensity, is partially also a global model parameter for the DI-SRM.

6.3.4 Computations

The introduced concept of mixing time determination is conceptually the same as the one presented in [113]. The main difference, besides the used GA implementation itself, is in the management of computations. In [113] optimisation steps were carried out in-line, i.e. the individuals within each population were calculated one after each other. Here, computations are carried out parallel with regards to both, the calculation of each individual and the calculation of a population. Such double-stage parallel computations have dramatically decreased the overall CPU time. Using the 28-species reaction mechanisms and the configuration A for the DI-SRM (Table 4.2), the parallel computation of a single engine cycle on 32 CPUs (Opteron 2378@2.4 GHz processors from the year 2008) takes approximately 7 s. Mean results are averaged over 100 cycles yielding 11.7 min for the complete run that is an effective cost of calculating single individual in GA nomenclature. Each population contains 10 individuals and these are also calculated in parallel. Therefore, simulating the assumed 15 populations requires approximately 3 hours. For the analysed operating point with single fuel injection, the solution fulfilling tolerances in Table 6.1 is found in the 10th population, yielding 1.95 hours for the complete simulation. Several tests performed on similar engine and using the same reaction mechanism indicate that the final solution is usually obtained after generating in between 7 and 12 populations. This corresponds to 1.3 to 2.3 hours for a complete calibration of a single engine operating point.

6.4 Parametrised DI-SRM for Engine Performance Studies

For an effective application of the DI-SRM for parameter studies or optimisation of engine performance parameters, the model should enable simulating of a full range of engine operating points (OPs) that are defined by speed and load. The in-cylinder pressure, RoHR and exhaust emissions should be accurately predicted by the DI-SRM in response to the changes of known engine design and operating parameters such as speed, amount of fuel injected, fuel injection rate, injection pressure, start of injection, injection duration, number of injections, EGR rate, valve timing strategy, combustion chamber geometry and several others [61]. This modelling challenge is illustrated conceptually in Fig. 6.9.

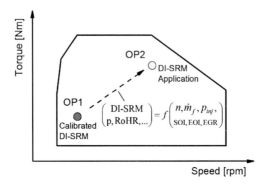

Figure 6.9: Simulations of engine performance maps using the DI-SRM

In Fig. (6.9), OP1 represents an operating point for which the mixing time and other DI-SRM parameters (see Section 4.4) are known from the calibration against the reference experimental or in some circumstances also 3D CFD data. In turn, OP2 represents an operating point for which there are no reference data, but for which the DI-SRM parameters must be known before simulations.

To enable the application as in Fig. 6.9, the modelled DI-SRM parameters must either be valid for the whole range of changes of engine operating parameters or they must be prescribed according to their changes. Such a prescription is one of the challenges in the applications of 0D engine models and is addressed in this section.

6.4.1 Overall Concept

The setup of DI-SRM is determined by the number of particles (N_P), the number of cycles (N_c), the time step (Δt) in the operator splitting loop, the stochastic heat transfer constant (C_h), the vaporisation rate and the mixing time (τ). Among these parameters, only the mixing time requires calibration. The remaining parameters can be estimated before simulations as presented in Section 4.4. This dependency is a foundation of the simulation process oriented at simulating a wide range of engine operating points and using only the information about the mixing time (Fig. 6.10).

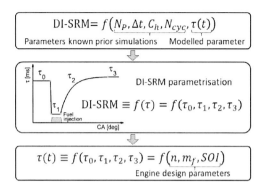

Figure 6.10: Parametrisation of the DI-SRM with engine design parameters

The mixing time depends mainly on in-cylinder flows that are influenced by engine speed, cylinder geometry, valve timing and fuel injection strategy. In this work, engine speed (n), mass of fuel injected (m_f), fuel injection rate $(\dot{m}_f)$ and its timing (SOI, EOI) as well as the number of injection events are the engine design parameters that influence the mixing time.[1] Changes of the design parameters imply the mixing time history varies between different engine operating points. As a result, the mixing times τ_0, τ_1, τ_2 and τ_3 (see Fig. 5.5) must be prescribed separately for each operating point. The engine design parameters are known before simulations and therefore, they can be used for the parametrisation of the mixing time model. The parametrisation describes the changes of τ_0, τ_1, τ_2 and τ_3 (see Eq. 5.7) in dependence to the changes of n, m_f and SOI.

6.4.2 Computational Setup

Simulations refer to Engine D specified in Table A.1 of Appendix A. This is a Diesel fuelled engine that was operated with single fuel injection. The DI-SRM was set-up according to configuration C in Table 4.2. The 28-species n-heptane mechanism (see Section 4.5.1) is used as Diesel surrogate. It was selected because of its low demand for CPU time that is beneficial during simulation method development.

---

[1]In the present chapter, henceforth and to simplify the description, engine design and operating parameters are commonly refereed to as the design parameters.

Measured exhaust NO_x and in-cylinder pressure data have been used as references for validating the method. The data refer to four different n and m_f values and five different SOI positions for each combination of n and m_f. At each operating point (OP) the engine was operated with 20% of EGR. The resulting engine matrix contains 80 points that are presented in Fig. 6.11.

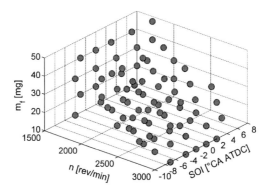

Figure 6.11: Matrix of reference experimental engine operating points

The engine under consideration is operated with single fuel injection and the setup of mixing time follows the reduced method presented in Section 6.3.3. The mixing time is described by only three parameters, namely, SOV, τ_1 and $t_{1/2}$. The remaining parameters, τ_0 and τ_3, are calculated based on the three previous. Among the engine design variables, n, m_f and SOI are selected for correlating with mixing time history. Hence, an overall setup of the mixing time can be expressed as

$$\tau(t) = f\big(SOV, \tau_1, t_{1/2}\big) = f(n, m_f, SOI).$$ (6.2)

The solution of Eq. (6.2) is denoted as DI-SRM parametrisation using tabulated mixing time and is discussed in the next subsection. The effect of fuel injection pressure, which strongly influences fuel and air mixing, is in Eq. (6.2) indirectly considered by having the effects due to operating parameters n, m_f and SOI. Each combination of these parameters has corresponding fuel injection rate profile that in some extent is a derivative of the fuel injection pressure. EGR has been excluded from the parametrisation since in general, it affects chemical processes, such as the formation of NO_x, and not physical processes such as mixing.

6.4.3 Mixing Time Tabulation

From the reference matrix of engine operating points shown in Fig. 6.11, five points were selected for the analysis of the impact of n, m_f and SOI changes on the mixing time history. The analysis is referred to as DI-SRM training and the operating points selected are henceforth referred to as *training operating points*. The training points were chosen to have at least three points on the engine map in the direction of n and m_f. With respect to SOI changes, all five available points were considered for each pair of n and m_f that were selected for model training.[1] Therefore, there are in total 25 training points out of 80 available. In Fig. 6.12, the operating points selected for the training of the DI-SRM are denoted by circles. Squares in that figure denote operating points that were used for the verification of the simulation method. Using one set of data for model training and the different set of data for model verification allows more comprehensive analysis of the modelling performance. For clarity reasons, Fig. 6.12 shows only main 16 points that differ by the mass of fuel injected and engine speed.

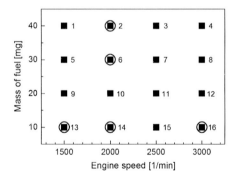

Figure 6.12: Selection of engine operating points for DI-SRM training (circles) and operating points for simulation method verification (squares)

---

[1]The choice of data from all five available SOI positions was imposed by high sensitivity of the mixing time modelling on the fuel injection process (see Section 5.4). However, most recent tests have indicated that the effect of SOI changes on the mixing time can be sufficiently accurately determined based on three different positions of SOI for each combination of n and m_f and for this particular engine configuration.

The mixing time model was calibrated separately for each training operating point using an automated procedure presented in Section 6.3. Six tracking parameters were selected to evaluate the quality of the obtained mixing time histories. Combustion progress is tracked based on CA_{10}, CA_{50}, CA_{90} and p_{max}. The parameters CA_{10} and CA_{90} are indicators of the occurrence of the SOC and EOC, respectively. Indicated mean effective pressure (IMEP)[1] is regarded as an overall indicator of engine performance. The formation of exhaust NO_x was controlled based on the concentrations computed at EVO.

The separate calibration of the DI-SRM at the selected engine operating points results in a set of single solutions for each parameter of the mixing time model. The results have a scatter character that can be treated as a manifold of solutions. To obtain a single solution for Eq. (6.2), first the mutual interactions were distinguished between the mixing time parameters and engine design variables. Subsequently, the obtained dependencies were approximated by polynomials. The dependencies and the setup of the representative mixing time history are presented in Fig. 6.13.

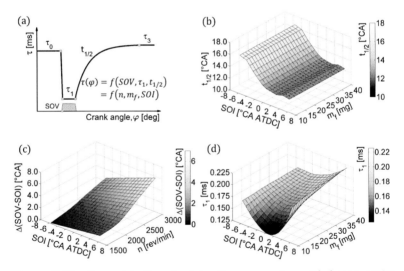

Figure 6.13: Parametrisation of mixing time parameters with known engine design variables; (a) representative mixing time history, (b) $t_{1/2} = f(SOI, m_f)$, (c) $\Delta(SOV\text{-}SOI) = f(SOI, n)$ and (d) $\tau_1 = f(SOI, m_f)$

---

[1]In this work, IMEP is calculated for the closed part of the cycle unless stated otherwise.

The surfaces in Figs. 6.13b to 6.13d were obtained through the interpolation in between the results obtained during the separate calibration of mixing time parameters at the training points. For engine operating parameters not considered during the training process, the values of SOV, τ_1 and $t_{1/2}$ were extrapolated.

In Fig. 6.13, instead of a correlation for the absolute values of SOV, the correlation for the difference between the SOV and the SOI is presented (see also Section 4.3). This is more general since it directly indicates the calibration range of the SOV and makes the correlation independent on the position of SOI. The position of SOV depends on n and SOI and does not depend on m_f. The changes of SOV as a function of SOI have linear character. The position of SOV is moved towards the later CA as the SOI is retarded. Delaying the SOI towards the expansion stroke implies the fuel injection occurs in larger volume, thus at the lower pressure and temperature. Lower temperature in the combustion chamber decreases the heat transfer between the injected fuel droplets and the surrounding gas. This slows down the fuel vaporisation process. As a result, the effective distance on the CA basis between the SOV and the SOI increases. For a given SOI, a higher engine speed implies that relatively more time passes until the vaporisation starts.

With respect to SOI and m_f changes, τ_1 is modelled by a third order polynomial. Mixing time τ_1 governs the mixing process during the fuel injection, vaporisation, air-fuel mixture preparation and early combustion phase. The intensity of these processes depends on the volume in which they occur. For very early or late fuel injection, mixing occurs in a large volume when compared to the volume at TDC. In consequence, the kinetic energy from fuel injection is faster dissipated at the extreme CA positions. During simulations, this corresponds to lower probability that all particles in the DI-SRM will mix with each other. It implies also longer τ_1 in these regimes. Contrary situation occurs for SOI defined at around TDC that implies using shorter mixing time τ_1. This interaction is reflected in the modelled history of τ_1 (Fig. 6.13d) that follows approximately a trajectory of the in-cylinder volume change.

The parameter $t_{1/2}$ is modelled by a fourth order polynomial with respect to SOI changes and by a constant value linear function with respect to m_f changes. The value of $t_{1/2}$ was optimised as the last component of the overall mixing time history. Hence, the dependency of $t_{1/2}$ upon the SOI, n or m_f is considered as the closure to the modelled dependencies for SOV and τ_1.

6.4.4 Performance of the Method

Combustion and NO$_x$ Emission

Results from Case 4 and Case 5 are analysed. They represent different engine load and speed (see Table B.1 of Appendix B). With respect to engine matrix in Fig. 6.12, they correspond to OP 14 and OP 8, respectively. The results from Case 4 are considered representative for points used for the development of dependencies between mixing time parameters and engine design variables in Fig. 6.13. In turn, Case 5 is an example of points excluded from the development of the dependencies in Fig. 6.13.

Figure. 6.14 shows comparisons between the simulated and the measured traces of in-cylinder pressure and rate of heat release. The simulated traces refer to two calculation methods. Results from the calibration of mixing time history refer to model training. Results obtained with the mixing time constructed based on the dependencies in Fig. 6.13 represent model validation.

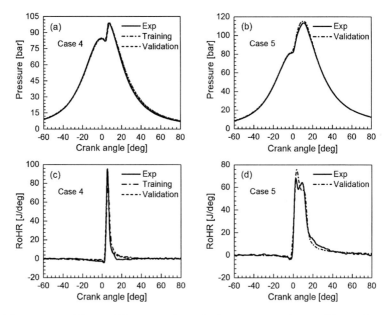

Figure 6.14: Measured histories of in-cylinder pressure and rate of heat release compared to the simulated traces from the training and validation of the DI-SRM

For Case 4, the simulated in-cylinder pressure and RoHR traces from the model training and validation agree between each other and match accurately the experimental traces. Similar accuracy has been obtained for Case 5 that was not involved in the training process.

The concentration of exhaust NO_x from the model training and validation is reasonably predicted for both cases. Slightly lower accuracy from the model validation is attributed to the reconstruction process of mixing time from the dependencies in Fig. 6.13. This aspect of modelling is discussed next.

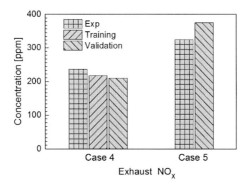

Figure 6.15: Measured exhaust NO_x compared to the computed values from the model training and validation for Case 4 and Case 5

Reconstructed Mixing Time and Locality of Mixture Formation

The dependencies for mixing time parameters in Fig. 6.13 are valid with a certain tolerance. Thus, the reconstructed mixing time may differ from the calibrated one as shown in Fig. 6.16 for Case 4. Between 1 °CA and 10 °CA ATDC, the mixing time traces from the model training and validation differ only slightly. After 10 °CA ATDC the difference between them becomes more noticeable, but its impact on the in-cylinder pressure and rate of heat release is negligibly small (Fig. 6.14). More visibly, the impact of the reconstructed mixing time is observed for exhaust NO_x in Fig. 6.15. The formation of NO_x is more sensitive than the in-cylinder pressure and rate of heat release to the changes of local in-cylinder temperature and equivalence ratio, which are governed by mixing intensity. The difference between the local in-cylinder temperature and equivalence ratio from the model training and validation is shown in Fig. 6.17.

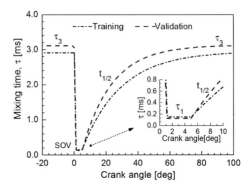

Figure 6.16: Mixing time from the model training and validation for Case 4

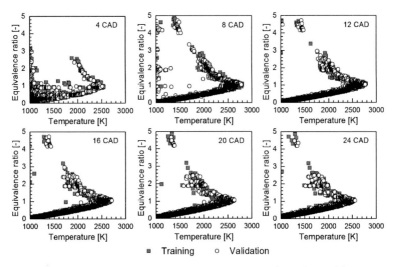

Figure 6.17: Local temperatures and equivalence ratios from the model training and validation at different instants during the cycle for Case 4

The overall trend of ϕ-T distributions is similar for both mixing time histories, but the difference in spread of the particles is also evident. It is expected that this spread has an influence on the local rates of formation of NO and NO_2, which results in different exhaust NO_x concentration. Aspects of NO_x formation are discussed in more detail in Section 7.3.

Results for the Complete Load-Speed Map

Based on the dependencies in Fig. 6.13, mixing time histories were derived for all operating points in Fig. 6.11. Next, the DI-SRM was applied for the simulation of in-cylinder performance parameters and exhaust NO_x at these points. Figure 6.18 and Fig. 6.19 present exemplary comparisons between the measured and the computed traces for in-cylinder pressure and rate of heat release, respectively. These results correspond to 16 pairs of engine speed and mass of fuel injected from Fig. 6.12 and for the middle position of SOI for each pair of speed and mass of fuel injected in Fig. 6.11. Overall, the computed traces of the in-cylinder pressure and rate of heat release rate follow accurately the corresponding measured counterparts.

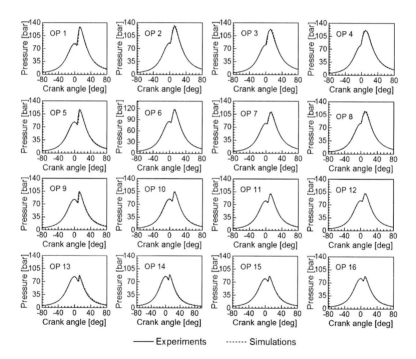

Figure 6.18: Computed and measured in-cylinder pressure traces for operating points depicted in Fig. 6.12 and for the middle position of SOI for each pair of m_f and n in Fig. 6.11

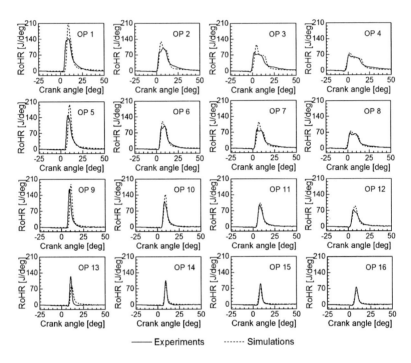

—— Experiments   ······· Simulations

Figure 6.19: Computed and measured RoHR traces for operating points depicted in Fig. 6.12 and for the middle position of SOI for each pair of m_f and n in Fig. 6.11

To provide an overview of the results from all 80 operating points in Fig. 6.11, scatter plots were created for the selected engine in-cylinder performance parameters such as IMEP, maximum pressure peak, exhaust NO_x and combustion progress that is tracked based on CA_{10}, CA_{50} and CA_{90}. The results are presented in Figs. 6.20 to 6.22. Furthermore, to indicate the quality of agreement between the measured and the simulated quantities, the coefficient of determination (R^2) was calculated (see, e.g., [129]) for each considered set of variables. Overall, the best correlation between the computed and the measured data is obtained for CA_{10}, CA_{50} and p_{max}, for which the R^2 coefficient exceeds 0.95 (Fig. 6.20). For CA_{90}, R^2 is 0.655 and the results are more spread than for CA_{10} and CA_{50}. For exhaust NO_x (Fig. 6.21a), R^2 is 0.70, but the correlation between the computed and the measured values is regarded as still acceptable. The highest deviation between the simulated and the experimental data is observed for high load operating points and most retarded SOI.

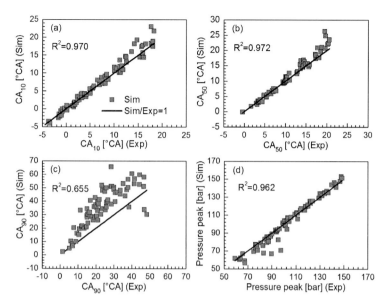

Figure 6.20: Prediction of the combustion progress and maximum pressure peak for all operating points from the matrix in Fig. 6.11

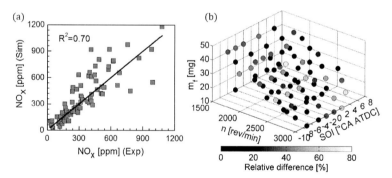

Figure 6.21: Exhaust NO_x for data points in Fig. 6.11; correlation between the computed and the measured data (a) and the accuracy of simulations (b)

Good agreement between the simulated and the measured in-cylinder pressure traces yielded also good agreement in IMEP. Figure 6.22 presents IMEP for each pair of n and m_f and for different angular positions of SOI such that (a) contains results for most early SOI and (e) for most late one in Fig. 6.11.

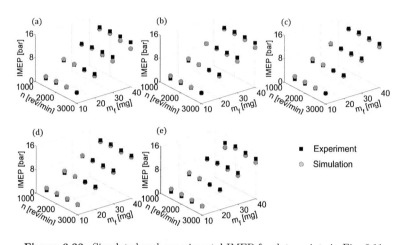

Figure 6.22: Simulated and experimental IMEP for data points in Fig. 6.11

Overall, good agreement between the simulated and the measured quantities over the wide range of engine operating points proves the plausibility of (1) the concept of representative volume-averaged mixing time model introduced in Chapter 5.2 and (2) the concept of mixing time parametrisation with known engine design parameters for predictive simulations of engine load-speed performance maps, which is introduced in the present section.

Computations

The computational cost of mixing time parametrisation increases as the number of operating points used to derive the dependencies for mixing time increases. Using configuration C (Table 4.2) of DI-SRM settings ($N_C=30$, $N_P=1000$, and $\Delta t=0.5$ °CA) and the 28 species n-heptane mechanism it is possible to calibrate the DI-SRM at a single operating point within approximately 4.6 hours. Once the mixing time has been determined, simulations for the same DI-SRM configuration take approximately 26 min (see Section 6.3.2, p.111). Here, calculations were carried out in parallel on 16 CPUs of the computer cluster equipped with AMD Opteron 2378 @2.4 GHz processors from the year 2008. Model calibration was carried out based on an automated method presented in Section 6.3.

Chapter 7

Simulation of Emission Formation and Fuel Effects

7.1 Introduction

The previous three chapters describe aspects of mixing time modelling and engine simulation process development for the DI-SRM. The emphasis was on further improvements of the DI-SRM to expand its potential for the simulation of combustion and emission formation in Diesel engines. In this chapter, three exemplary applications of the improved DI-SRM for the simulation of Diesel engines are presented. First, an application of the DI-SRM for simulating the performance of a Diesel engine fuelled with different fuel blends is demonstrated. Next, an analysis of kinetics effects in simulating NO_x formation is carried out. Subsequently, aspects of 0D modelling of soot formation are presented.

7.2 Fuel Effects Under Diesel Conditions

The efficiency of engine operation results from the interactions between the fluid dynamic, thermodynamic and chemical processes. As a consequence, Diesel engine operating parameters, such EGR, fuel injection characteristic, intake temperature and pressure, must match with fuel properties such as heat release characteristic, cetane number index, lower heating value or the content of aromatic, sulphur and other components. These properties are decisive for autoignition, combustion rate and emission formation processes. The fuel properties vary over the world market fuel depending on the legislation, production processes or crude oil composition. Therefore, the changes of fuel properties also require consideration during the engine development. Investigating the performance of different fuel blends under engine conditions is also needed while designing new fuel compositions or for the development of surrogate fuels for engine modelling (see, e.g., [41, 123]). Such investigations can be supported by the PDF-based DI-SRM thanks to detailed chemistry consideration. In this section an exemplary application of the DI-SRM for the analysis of fuel effects under Diesel conditions is presented.[1]

7.2.1 Simulation Method and Surrogate Fuels

The numerical model consists of data, software and application rules as depicted in Fig. 7.1. The module *Physics* corresponds to a set of submodels for the simulation of engine in-cylinder processes of IC engines. Here, the DI-SRM is used. The module *Chemistry* denotes a set of reaction mechanisms, species thermodynamics and physics for various hydrocarbon fuels. This database enables the defining of single or multicomponent surrogates of actual fuels. The module *Optimisation* is a general purpose genetic algorithm to control, automate and speed-up the overall engine simulation process. Depending on the application rules, the modelling concept in Fig. 7.1 can be applied for the analysis of various aspects of engine-fuel interactions. In the current work, it has been used to estimate the influence of differently composed fuel blends on Diesel engine performance parameters.[2]

Calculations were carried out for Case 2 described in Table B.1 of Appendix B. The engine was operated at low speed and part load and with single injection of n-heptane. The setup of the DI-SRM follows configuration A presented in Table 4.2. The results presented in Section 5.4.4 (Fig. 5.9 and Fig. 5.10) are basis for the simulations in this section. In terms of in-cylinder pressure and exhaust emission, they were verified against the experimental data from the n-heptane fuelled engine. For simplicity, the baseline results are referred to in this section as Case A.

---

[1]This section is based on an article [114] published in the Proceedings of the 8th International Conference on Modelling and Diagnostics for Advanced Engine Systems (COMODIA), Fukuoka, Japan, July 23-26, 2012.

[2]Based on the work here presented, the modelling concept in Fig. 7.1 has been recently commercialised and released as a part of the software package LOGEengine [87].

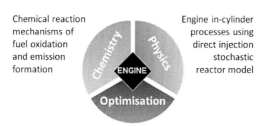

Chemical reaction
mechanisms of
fuel oxidation
and emission
formation

Engine in-cylinder
processes using
direct injection
stochastic
reactor model

Genetic algorithm for process
control, optimisation and automation

Figure 7.1: Modelling framework for the analysis of engine-fuel interactions

Four different fuels have been selected for the comparison to each other, namely, pure n-heptane, iso-octane-doped n-heptane, toluene-doped n-heptane and pure n-decane. They are frequently recommended as surrogate components of real Diesel fuel (see, e.g., [41]). The main properties of these fuels are given in Table 7.1. For simplicity, later in this section, the fuels are referred to as Fuel A, Fuel B, Fuel C and Fuel D, respectively. The corresponding DI-SRM setups are denoted as Case A, Case B, Case C and Case D.

Table 7.1: Basic characteristic of Diesel and surrogate fuels

Fuels	CN[-][a]	LHV[MJ/kg][b]
Diesel	40-56	43.20
A – n-heptane	56	44.60
B – n-heptane/iso-octane (0.8/0.2, vol.-based)	47	44.56
C – n-heptane/toluene (0.8/0.2, vol.-based)	45	43.80
D – n-decane	77	44.24

[a] CN based on [33, 41, 103].
[b] Lower Heating Value (LHV) based on [25, 33].

Reaction mechanisms for the selected fuels A, B, C and D are described in [173], [138], [1] and [63], respectively. These references provide a detailed description of the mechanisms and discuss their validation against a variety of experimental data. The n-heptane mechanism is also briefly described in Section 4.5.2 (the 121-species mechanism).

7.2.2 Chemical Effects of Fuel Composition

To estimate the impact of chemistry on the combustion rate and exhaust emission, the same setup of the DI-SRM, i.e. SOI, EGR, intake pressure, intake temperature and mixing time, was used for the simulation of all four fuels listed in Table 7.1. The setup has been taken from the baseline Case A. The difference in LHV between the fuels has been compensated be adjusting the mass of injected fuel to ensure the same energy released during the combustion process.

In Fig. 7.2, the in-cylinder pressure and RoHR histories for the investigated four fuels are presented. The shortest combustion duration and highest peak of the in-cylinder pressure and RoHR is obtained with pure n-decane (Fuel D) and pure n-heptane (Fuel A). Including 20% of iso-octane in the mixture with n-heptane (Fuel B) delays slightly the start of combustion when compared to pure n-heptane. The combustion occurs with a lower rate and the duration is longer. The CA_{50} occurs later in the cycle and IMEP decreases (Fig. 7.3). As a result, the maximum pressure peak and gradient also decrease. Similar behaviour is observed for the toluene-doped n-heptane. However, the impact of toluene on the mixture and thus the rate of heat release profile, is more noticeable than that of iso-octane.

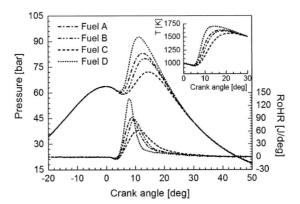

Figure 7.2: Histories of the in-cylinder pressure, temperature and RoHR for Fuels A, B, C and D defined in Table 7.1 (DI-SRM setup adopted from Case A)

The differences between the fuels in terms of in-cylinder pressure, RoHR and temperature result mainly from the differences in the chemical ignition delay time and the nature of the rate of chemical energy release. The setup of the DI-SRM is the same for all four cases, hence the impact of the physical engine processes, such as mixing or fuel vaporisation, is here excluded.

The inclusion of iso-octane or toluene in the mixture with n-heptane deteriorates the autoignition since the CN decreases (Table 7.1). This tendency finds confirmation in the simulated pressure and RoHR histories in Fig. 7.2 and corresponding results for IMEP, CA_{50} and engine exhaust emissions in Fig. 7.3.

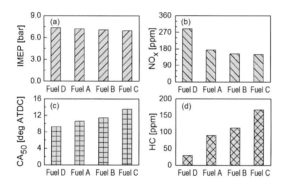

Figure 7.3: IMEP, CA_{50}, exhaust HC and NO_x for fuels from Table 7.1 and for the same engine operating parameters such as speed, load and fuel injection strategy

The highest concentration of exhaust HC was obtained for toluene-doped n-heptane (Fuel C), which has the longest combustion and lowest maximum pressure peak. The lowest concentration of exhaust HC is observed for pure n-decane (Fuel D) that reaches the highest peaks in the rate of heat release, pressure and temperature. The combustion process is enhanced and consequently less unburned HC are found at EVO. Contrary effects are observed for exhaust NO_x.

A very small difference in exhaust NO_x is noted between the Fuel B and the Fuel C (Fig. 7.3b). At the same time, however, the corresponding RoHR and in-cylinder pressure histories differ significantly. This issue is discussed separately in Section 7.3.

7.2.3 Engine Performance for Various Fuels

Varying the composition of the fuel blend while holding engine parameters constant, as in the previous subsection, allows for the separating from each other the physical and chemical processes occurring in the engine cylinder. However, no conclusion can be drawn on fuel quality under engine conditions. To overcome this limitation, engine design parameters, such as fuel injection strategy or EGR, must be set by considering the physical and chemical properties of the blends. As a computational test case, the performance of Fuel C (toluene-doped n-heptane) with reference to the baseline Fuel A (pure n-heptane) is analysed (see Fig. 7.2 and Fig. 7.3).

For the engine setup, which was optimal for the reference Fuel A, the engine operation for Fuel C is less efficient (Fig. 7.4 and Fig. 7.5). The IMEP is reduced, autoignition is delayed, combustion duration is increased and CA_{50} moves towards the expansion stroke.

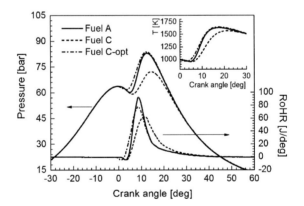

Figure 7.4: Histories of the in-cylinder pressure, temperature and RoHR for n-heptane (Fuel A) compared to the results from a blend of n-heptane and toluene (Fuel C) before and after the optimisation of SOI (Fuel C-opt)

To improve engine operation for Fuel C, the SOI has been optimised. The value of IMEP and the position of CA_{50} from Fuel A have been selected as optimisation targets. After optimisation, the SOI for Fuel C is advanced by about 2.0 °CA (Fuel C-opt). This advancement compensates high resistance to autoignition of toluene. As a result, the IMEP is increased to 7.27 bar and exceeds the value of the reference fuel (7.21 bar). The combustion duration is longer for the Fuel C although they both have very similar IMEP.

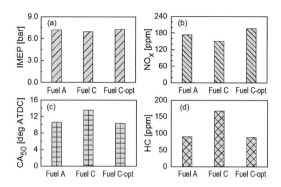

Figure 7.5: IMEP, CA$_{50}$, exhaust HC and NO$_x$ from n-heptane (Fuel A) compared to the results from a blend of n-heptane and toluene (Fuel C) before and after the optimisation of SOI (Fuel C-opt)

The RoHR history is also milder for the Fuel C, resulting in lower pressure gradients. Finally, despite the inclusion of toluene, the usage of Fuel C results in lower HC and slightly higher exhaust NO$_x$ in its optimal operation point.

Overall, in comparison to the results in Section 7.2.2, the results in Section 7.2.3 ensure a more plausible evaluation of the performance of Fuel A and Fuel C under engine conditions.

7.3 Kinetic Effects in Simulating NO$_x$ Formation

Thanks to the detailed treatment of fuel oxidation and emission formation chemistry by the DI-SRM, important intermediate or precursory species can be tracked. The reaction pathways or fuel properties, such as resistance to self-ignition, can be investigated. As a result, differences between variously composed fuels can be analysed in detail and under engine conditions. This section[1] demonstrates an application of the DI-SRM for the analysis of NO$_x$ formation that besides soot (discussed in next section) is most important pollutant from Diesel engines.

---

[1]This section is based on an article [109] published in Berichte zur Energie- und Verfahrenstechnik (BEV) 13.1, 337-346, ISBN 3-931901-87-4, 2013.

7.3.1 Baseline Results

Exhaust NO_x from the blend of iso-octane and n-heptane (Fuel B) and the blend of toluene and n-heptane (Fuel C) is analysed. The results have been extracted from Section 7.2.2 and are repeated in Fig 7.6 for readability of the subsection.

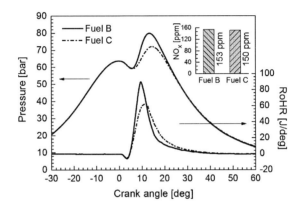

Figure 7.6: Simulated in-cylinder pressure, RoHR history and exhaust NO_x for n-heptane blended with iso-octane (Fuel B) and with toluene (Fuel C) for the same engine speed, load and fuel injection strategy in the DI-SRM

Despite the significant differences in the in-cylinder pressure and rate of heat release between Fuel B and Fuel C, the engine cycle for both blends result in almost the same exhaust NO_x, which on mass basis is 153 ppm and 150 ppm for Fuel B and Fuel C, respectively. Both blends use the thermal model of NO formation from [141]. These results are analysed in the next two subsections using the information from the reaction kinetics and locality of emission formation.

7.3.2 Temporary Evolution of NO_x

Before approximately 55 °CA ATDC, the cumulative concentration of NO is higher for Fuel B (Fig. 7.7). It can be explained by the differences in the mean gas temperature (Fig. 7.8). Until 30 °CA ATDC, the temperature for Fuel B is higher, but then it drops faster than for Fuel C. As a result, at approximately 55 °CA ATDC the concentration of NO is similar for both fuels. Then it

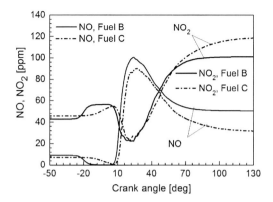

Figure 7.7: Simulated crank angle resolved formation of exhaust NO and NO_2 from n-heptane blended with iso-octane (Fuel B) and with toluene (Fuel C)

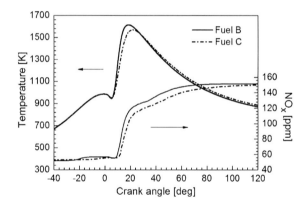

Figure 7.8: Simulated exhaust NO_x and mean in-cylinder temperature for n-heptane blended with iso-octane (Fuel B) and with toluene (Fuel C)

starts freezing for Fuel B. At the same time, for Fuel C the concentration of NO drops further and the concentration of NO_2 increases towards EVO. The increase of the concentration of NO_2 occurs on the expense of NO via the known reaction with HO_2: $NO+HO_2 \rightarrow NO_2+OH$ [128, 145, 146]. As a result, the overall mass-based concentration of NO_x is increased for Fuel C since NO_2 has a higher molar mass than NO.

7.3.3 Local Results in ϕ-T Space

Further insight into the formation of NO_x is gained from the distributions of local equivalence ratio and temperature in Fig. 7.9 and corresponding PDF distributions of mole fraction for NO and NO_2 shown in Fig. 7.10.

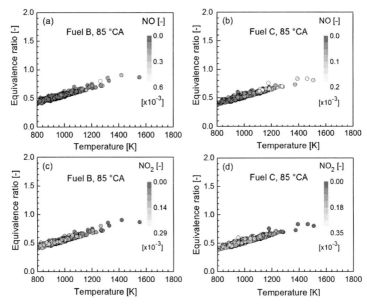

Figure 7.9: Mole fractions of NO and NO_2 for n-heptane blended with iso-octane (Fuel B) and with toluene (Fuel C) in ϕ-T space

The results presented in Fig. 7.9 and Fig. 7.10 refer to 85 °CA ATDC at which the combustion process has ended (see Fig. 7.6) and there is already a substantial difference in NO and NO_2 concentration between the blends (Fig. 7.7). In Fig. 7.9, the majority of particles are in a temperature range between 800 K and 1500 K. Overall, the particle-based content (mole fraction) of NO reaches lower values for Fuel C than for Fuel B. A contrary situation is observed for NO_2. Confirmation of this tendency is observed in Fig. 7.10, which indicates a slightly more homogeneous distribution of NO concentration and less homogeneous NO_2 for Fuel C when compared to Fuel B. This results in a higher concentration of NO_2 at EVO for Fuel C, which also increases NO_x concentration. As a consequence, NO_x is similar for both fuels.

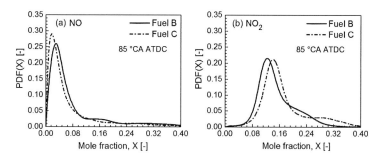

Figure 7.10: PDF distributions of mole fraction of NO and NO_2 for n-heptane blended with iso-octane (Fuel B) and with toluene (Fuel C) at 85 °CA

7.4 Aspects of Simulating Soot for Diesel Engines

The prediction of soot from Diesel engines remains a challenging task. Under conditions with turbulence and chemistry interactions, the modelling requires consideration of different time scales of the chemical and physical processes involved in soot formation and oxidation and their locality in space.

This section[1] demonstrates an application of the DI-SRM as diagnostic tool for a plausibility study for assumptions made for the flamelet soot source term model that is frequently used ([6–8, 104, 105, 165]) in 3D CFD modelling of soot formation in Diesel engines. The modelling assumes that gas-phase species for the calculation of the soot source terms (particle inception, surface growth, fragmentation and oxidation) can be taken from a stationary flamelet library and the time scale for soot formation is too long to allow the application of transient interactive flamelet models. In this context, the benefit of the DI-SRM is the calculation of gas-phase chemistry and soot formation through detailed kinetics. Furthermore, using PDF distributions of scalar properties, such as species concentrations, temperature, reaction progress and mixture fraction, the local effects in the combustion chamber under Diesel conditions can be captured. On the other hand, the mixing time for the DI-SRM can be verified by the 3D CFD data.[115]

---

[1]This section is derived in part from an article [115] published in Combustion Science and Technology on 30 September 2014, available online: http://wwww.tandfonline.com/DOI: 10.1080/00102202.2014.935213.

Computations in this section refer to Case 1 described in Table B.1 of Appendix B, which was also used in Section 5.4. Setup of the DI-SRM follows configuration C from Table 4.2. Gas-phase reactions are described by the 121-species n-heptane mechanism that is discussed in Section 4.5.

7.4.1 Locality of Soot Formation

Figure 7.11 depicts mole fractions of C_2H_2, OH, NO and CO in mixture fraction space at 15 °CA and 30 °CA ATDC. These CAs correspond to soot formation and oxidation phases respectively (see Fig. 7.15).

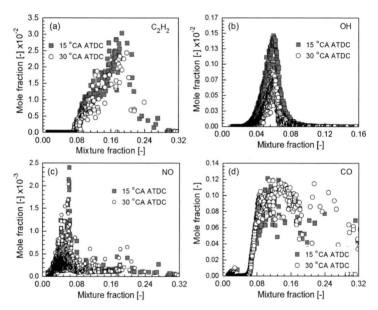

Figure 7.11: Mole fractions of C_2H_2, OH, NO and CO in mixture fraction space calculated by the DI-SRM at 15° CA and 30° CA ATDC

The species displayed in Fig. 7.11 are selected on the basis of their formation and consumption chemistry occurring at different time scales. C_2H_2 is the most important growth species while OH is the most important for soot oxidation. The flamelet structure for these two species is very well established by the DI-SRM calculation (Fig. 7.11). NO and CO react on the slowest chemical time scale and present a wider spread of data points. For NO the flamelet approximation might be questionable since a large variation of the NO mass fraction exists for each coordinate in mixture fraction space and there is no simple relation between mixture fraction and NO mass fraction. Hydroxyl is the fastest reacting species and shows a perfect one dimensional structure. As opposed to that, the calculated first moment of the soot particle size distribution function exhibits more spread and does not manifest a one dimensional flamelet-like structure (Fig. 7.12).

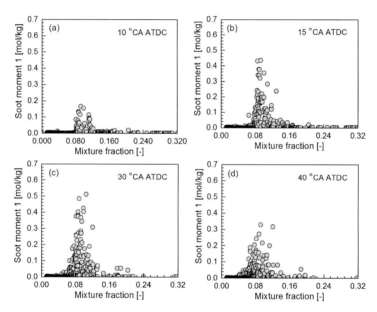

Figure 7.12: The first moment of the soot particle size distribution function calculated by the DI-SRM at 10 °CA, 15 °CA, 30 °CA and 40 °CA ATDC

For the same mixture fraction coordinate, it is noticed that a wide dispersion of the soot moments exists. Hence, it can be more accurate to use the rates of soot formation together with flamelet models than to calculate the soot concentrations through the interactive flamelet model. This is partially proved by looking at the calculated soot source terms for particle inception and surface growth (Fig. 7.13a,b) and oxidation (Fig. 7.13c,d); the flamelet structure is better established for the source terms than for the calculated first moment of the soot particle size distribution function (Fig. 7.12).

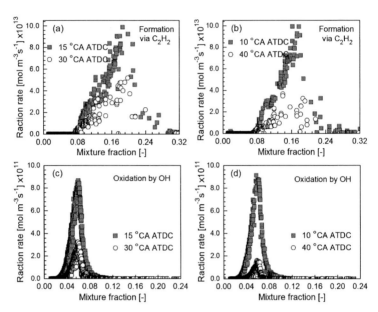

Figure 7.13: Soot formation via C_2H_2 and oxidation via OH in mixture fraction space, and calculated by the DI-SRM at 15 °CA and 30 °CA ATDC

The calculated soot source terms for particle inception, surface growth and oxidation (Fig. 7.13) remain the flamelet structure. However, it is also noted that the distribution of the source terms is more spread on the fuel lean conditions. To analyse these results in more detail, the progress variable (C) of the combustion process has been added to the analysis. For each particle the progress variable was calculated based on the enthalpy of formation at the reference temperature as in [83].

In Fig. 7.14 the reaction rates (r) are presented for soot growth by C_2H_2 and oxidation by OH in Z-C space, where Z denotes mixture fraction. The results refer to 15 °CA and 30 °CA ATDC that correspond to soot formation and oxidation (see Fig. 7.15). The results indicate that the reaction progress is responsible for some of the observed spread in the mixture fraction coordinate below the stoichiometric condition (0.064 for n-heptane). More spread towards the later oxidation phase in Fig. 7.14b is attributed to the lower temperature, which prevents further oxidation. This is caused by the cooling effect of the expansion (pressure work) and wall cooling (heat transfer), which increases during the expansion cycle and increases the spread of temperature.

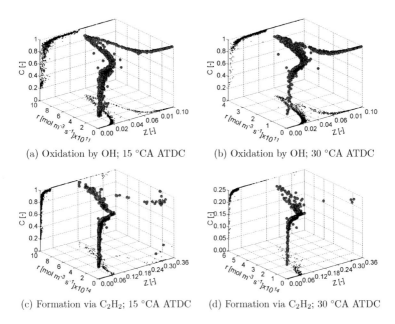

(a) Oxidation by OH; 15 °CA ATDC (b) Oxidation by OH; 30 °CA ATDC

(c) Formation via C₂H₂; 15 °CA ATDC (d) Formation via C₂H₂; 30 °CA ATDC

Figure 7.14: Scatter plots for the rates of soot oxidation via OH (a, b) and soot formation via C_2H_2 (c, d) calculated at 15 °CA and 30 °CA ATDC

7.4.2 Practical Aspects of Soot Modelling

Experimentally obtained soot concentration at EVO and profiles of soot formation and oxidation calculated by the 3D CFD model are used as reference for DI-SRM simulations. Both the reference experimental soot data and the corresponding 3D CFD calculations are based on [104]. In the 3D CFD model, soot formation and oxidation was calculated based on the flamelet library of soot sources. The same chemical schemes for gas-phase (n-heptane) and PAH chemistry in the soot model was used in the DI-SRM and the 3D CFD model.

Figure 7.15 compares the profiles of soot formation and oxidation from the DI-SRM with the 3D CFD computations and the experimental data at EVO. Soot surface growth and soot consumption via oxidation by OH have been calibrated in the DI-SRM to obtain correct crank angle resolved soot formation and oxidation. This corresponds to scaling of the soot source terms in the 3D CFD model [104]. For both the DI-SRM and 3D CFD model, the soot mass is normalised to the mass of fuel.

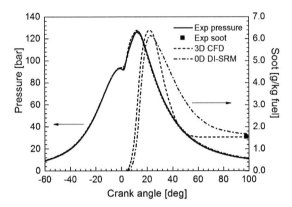

Figure 7.15: Histories of soot formation/oxidation and in-cylinder pressure from the DI-SRM and 3D CFD compared to the measured data

The general characteristic of the soot formation process are very similar for both models but the soot oxidation is slightly different. At first, soot oxidation appears to be faster in the CFD simulations. Towards the end of the cycle, soot oxidation is frozen in CFD while it continues in the DI-SRM. The quenching of the soot oxidation is explained by missing tabulation entries for the expansion stroke. The direct chemistry evaluation in the DI-SRM shows that soot oxidation continues at the late phase in the engine cycle. Tabulation methods for the expansion stroke of piston engines are under evaluation in the literature [68].

Chapter 8

Summary and Concluding Remarks

A probability density function (PDF)-based modelling has been applied for the calculation of combustion, emission formation and fuel effects in Diesel engines. The modelling combines a zero-dimensional (0D) direct injection stochastic reactor model (DI-SRM) of engine in-cylinder processes with complex reaction mechanisms of the oxidation process and emission formation from hydrocarbon fuels. Emphasis has been placed on aspects of mixing time modelling that is crucial for the overall performance of the DI-SRM.

8.1 General Findings

The PDF-based 0D DI-SRM is capable of predicting the in-cylinder performance of direct injection Diesel engines. The crank angle dependent volume-averaged mixing time improves the modelling of interactions between turbulent flow and chemistry. Using PDF distributions, the mixing time enables mapping of the 3D effects of in-cylinder flows, fuel injection and spray formation on the in-cylinder mixture composition in the 0D computational domain. As a result, the modelling of local inhomogeneity of the mixture in terms of equivalence ratio and temperature is improved, yielding enhanced accuracy of the simulation of in-cylinder pressure, rate of heat release and exhaust NO_x, HC and soot.

Parametrisation of the mixing time with known engine design parameters, such as speed, load and fuel injection strategy, makes the model independent on prior 3D CFD results. It enables the simulation of engine load-speed maps for a variety of operating parameters. The detailed chemistry consideration enables the investigation of differently composed fuels under engine conditions The 0D modelling framework makes the DI-SRM computationally efficient. In turn, the PDF-based formulation provides an insight in to the locality of mixture formation in the gas-phase and emission formation.

8.2 Conclusions

DI-SRM Tailor-Made for Diesel Engines

Mixing time is the only modelled parameter of the DI-SRM. Model numerical parameters, such as the number of particles (N_P), the number of consecutive cycles (N_C) and the global calculation time step (Δt) can be assumed before simulations. The results obtained indicate that $N_P=1000$, $N_C=30$ and $\Delta t=0.5$ °CA can be considered as a most general setup. It is relevant for the prediction of integral in-cylinder properties, such as pressure and RoHR, and local ones such as equivalence ratio, temperature and in consequence also exhaust emissions. The setup ensures acceptable computational cost for a given size of the reaction mechanism. The calibration of mixing time and the quality of the reaction mechanism representing Diesel surrogates are crucial for the overall performance of the DI-SRM.

Volume-Average Representative Mixing Time

The crank angle dependent and volume-averaged mixing time, which distinguishes between different time scales during the engine compression, fuel injection, spray formation, combustion and expansion phase, improves the overall performance of the DI-SRM under Diesel conditions. Accurate prediction of engine exhaust emissions is a main benefit of the proposed model over the approach employing a constant value mixing time.

Using the turbulent mixing time from 3D CFD and calibrating only the mixing time constant, gives similar results as the application of constant value mixing time. The inefficiency of such a modelling indicates the necessity for further consideration. Possible reasons for the observed inaccuracy in predicting the in-cylinder pressure and rate of heat release can be due to simplifications made while transforming 3D physical process into 0D modelling framework. The transformation goes via the turbulent mixing time that is not conditional with respect to the mixing process it governs. Further shortcomings can be the realisation of the mixing process itself and the negative impact of the missing modelling of near-wall processes such as fuel deposition on the wall or flows into and out of crevices. The effect of these processes on mixture composition and specifically HC emission is partially achieved by calibrating the mixing time. Hence, mixing time is also a global model parameter for the DI-SRM. Besides governing the intensity of mixing it must, in some extent, compensate the missing effects of 3D geometry on engine in-cylinder processes. Similarly, mixing time may also compensate possible inefficiency of reaction mechanisms. In this context, to improve the accuracy of the DI-SRM results, it is necessary to also model/parametrise the shape of the mixing time history as it has been proposed in this work. The resulting mixing time history is referred to as *parametrised mixing time model*.

Usually, for engines operated with single fuel injection, it is possible to describe accurately the mixing time by the parametrisation with just three constants. The constants govern the intensity of mixing before any fuel has been injected, during the fuel injection and vaporisation and during the mixing controlled combustion, towards exhaust valve opening. The proposed mixing time model is general. It has been developed for engines operated with single fuel injection, but it can also be extrapolated to engines operated with multiple fuel injection. This is obtained by superimposing mixing time history that is characteristic for a single fuel injection to other injection events.

In some cases it is possible to model mixing time for multiple fuel injection operation in a similar manner as for single fuel injection. Model parameter studies have indicated that for an engine operation with a short gap between the pilot and the main injection and with a small fraction of the fuel injected during the pilot, the mixing time during the two fuel injection events can be approximated by just one linear function. Such as simplification does not deteriorate the overall performance of the DI-SRM, but it significantly simplifies the overall simulation process and reduces its computational cost.

Automated Method of Mixing Time Determination

The determination of the modelled mixing time can be automated through an extended heat release analysis. The constants governing the mixing time history are optimised using a genetic algorithm to match the simulated in-cylinder pressure and exhaust emissions with the experimental counterparts. For engines operated with single fuel injection, a minimum three pressure targets must be defined. The targets should be defined around the start of combustion, maximum pressure peak and the end of combustion.

For multiple fuel injection, the number of pressure targets must be increased respectively. For an accurate prediction of exhaust emissions, such as unburned HC, NO_x or soot, they should also be considered as optimisation targets due to high sensitivity of pollutants formation on the locality of mixing process. The unburned HC and soot are more sensitive to the mixing time changes than NO_x. Usually, exhaust NO_x can be accurately predicted if the in-cylinder pressure and RoHR are accurately simulated and if the properties of actual fuel are accurately represented by the surrogate fuel that is used during simulations. Overall, through the calibration of mixing time exhaust NO_x and HC can be matched with 5% tolerance and in-cylinder pressure with 1% if sufficiently accurate surrogate fuel is applied.

The automated method of mixing time determination via the GA-based extended heat release analysis as proposed in this work has been recently commercialised and released within a software package LOGEengine [87].

Parametrised Mixing Time for Engine Performance Mapping

The history of mixing time can be parametrised with known engine design parameters such as speed, mass of fuel injected, injection rate and timing. The parametrisation of mixing time and an exact treatment of the non-linearity of reaction kinetics that influence the local rates of heat release and pollutants formation allow for the extrapolation of engine performance parameters beyond the operating points used for model calibration. In this work, the method was successfully applied for the prediction of engine in-cylinder pressure, rate of heat release and exhaust NO_x at wide range of engine speeds and loads. Most accurate correlations between the measured and the computed quantities across all operating points were obtained for CA_{10}, CA_{50}, p_{max} and IMEP for which the R^2 coefficient exceeds 0.95. Slightly lower accuracy was obtained for CA_{90} (R^2=0.65) and exhaust NO_x (R^2=0.70). Lower accuracy for exhaust NO_x is attributed to rather simple n-heptane mechanism, which was used because of its computational efficiency that is beneficial during simulation method development. Further impact has the difference between the n-heptane fuel used during simulations and the real Diesel fuel used during experiment. Overall, good agreement between the simulated and the measured quantities across all operating points indicate the method is general. An improvement of the prediction of exhaust NO_x is expected if more accurate fuel surrogate – preferable a multicomponent one – is used to represent properties of actual Diesel fuels. More complex Diesel surrogate should also enable an accurate simulation of exhaust HC and soot.

Simulation of Fuel Effects and Soot Formation

The application of complex reaction mechanisms in the DI-SRM enables the detailed investigation of fuel oxidation and emission formation. The formation of important intermediate or precursory species and reaction pathways can be tracked. Fuel properties, such as resistance to self-ignition, are accurately accounted for. As a result, differences between variously composed fuels can be analysed under engine conditions. In this work, these features of the DI-SRM have been used for the analysis of NO_x formation from the mixtures of n-heptane blended with iso-octane and toluene. The results obtained show that at a certain temperature window, during the expansion phase, there can be a substantial increase of NO_2 formation due to the oxidation of NO with HO_2. As a result, the total mass-based concentration of NO_x from the blend of n-heptane with toluene reaches almost the same level as the blend of n-heptane with iso-octane though the overall combustion rate and in-cylinder pressure are significantly higher for the latter mixture. Thus, the importance of detailed formulation of fuel surrogates is highlighted.

Thanks to detailed chemistry consideration, the DI-SRM can support 3D CFD modelling of emission formation, which frequently is performed in a simplified manner due to high cost of computations. In this work the DI-SRM was employed to investigate the applicability of the flamelet model for the combustion process in 3D CFD simulations and to elucidate the limitations of the interactive flamelet model when calculating emission formation. Species reacting on short time scales show profiles that agree very well with the flamelet theory. In turn, emissions like NO and soot show more dispersion. The dispersion observed for NO and soot casts doubt on the methodology of calculating emissions through the interactive flamelet model. One possible alternative is to apply source term libraries that are not limited to a specific combustion model and demand low CPU usage since chemistry does not need to be evaluated at runtime. More spread of the soot source terms for particle inception, surface growth and oxidation on the fuel lean conditions in the mixture fraction coordinate may be due to the reaction progress. The spread towards the later oxidation is attributed to the lower temperature that prevents further oxidation. This is caused by wall cooling and the cooling effect of the expansion; the engine enthalpy losses resulting from pressure work and heat transfer lead to an increase spread of temperature.

8.3 Recommendations for Future Work

The studies carried out in this work have indicated high sensitivity of the computational cost of DI-SRM simulations to the size of the reaction mechanism. Employing tabulated chemistry, similarly as in 3D CFD simulations, has potential to eliminate this drawback. Furthermore, the reduction of computational cost by tabulated chemistry would allow one to use more complex multicomponent Diesel surrogates that ensure closer match to the properties of actual Diesel fuels.

The modelling of mixing process is the main challenge of the PDF method. Investigating the performance of other existing mixing models, such as modified Curl or EMST, is needed to further expand the knowledge on the modelling of chemistry and flow interactions using the DI-SRM under Diesel conditions.

The dependencies obtained between the engine design variables, such as speed, start of injection and mass of fuel injected, and the parameters governing mixing time history seem general. Hence, to simplify the simulation process, the dependencies obtained for mixing time parameters could be directly optimised instead of the optimisation of mixing time parameters.

At present, Diesel engines are usually operated with multiple fuel injection strategy. Therefore, the modelling of mixing time history for fuel injection with more than two injection events and different timing and mass distribution between them should also be accounted for. Due to complexity of such an operation, it would be beneficial to correlate or deduce an overall history of the mixing time directly from the fuel injection data and spray modelling. This could be based on the k-epsilon turbulence model that enables the estimation of mixing time scale from the changes of the turbulent kinetic energy and its dissipation. Such an approach will not eliminate the necessity for calibration either. However, it should simplify the determination of the overall shape of the mixing time for complex multiple fuel injection strategies. The calibration in stead, could focus and suitable scaling of the obtained mixing time histories to match simulations with measurements.

References

[1] Ahmed, S., Mauss, F., and Zeuch, T. The Generation of a Compact n-Heptane / Toluene Reaction Mechanism Using the Chemistry Guided Reduction (CGR) Technique. *J. Phys. Chem.*, 223:551–563, 2009.

[2] Akihama, K., Takatori, Y., Inagaki, K., Sasaki, S., and A.M., D. Mechanism of the Smokeless Rich Diesel Combustion by Reducing temperature. *SAE Paper 2001-01-0655*, 2001.

[3] Amnéus, P. *Homogeneous Ignition- Chemical Kinetic Studies for IC- Engine Applications*. PhD thesis, Lund University, 2002.

[4] Amnéus, P., Mauss, F., Vressner, A., and Johansson, B. NOx and N2O Formation in HCCI Engines. *SAE Paper 2005-01-0126*, 2005.

[5] Asay, R., Svensson, K., and Tree, D. An Empirical, Mixing-Limited, Zero-Dimensional Model for Diesel Combustion. *SAE Paper 2004-01-0924*, 2004.

[6] Bai, X., M. Balthasar, M., Mauss, F., and Fuchs, L. Detailed Soot Modeling in Turbulent Jet Diffusion Flames. *Proc. Comb. Inst.*, 27(1):1623–1630, 1998.

[7] Balthasar, M. *Detailed Soot Modelling in Laminar and Turbulent Reacting Flows*. PhD thesis, Lund University, 2000.

[8] Balthasar, M., Heyl, A., Mauss, F., Schmitt, F., and Bockhorn, H. Flamelet Modeling of Soot Formation in Laminar Ethyne/Air-Diffusion Flames. *Proc. Comb. Inst.*, 26(2):2369–2377, 1996.

[9] Balthasar, M., Mauss, F., Knobel, A., and Kraft, M. Detailed Modeling of Soot Formation in a Partially Stirred Plug Flow Reactor. *Combust. Flame*, 128(4):395–409, 2002.

[10] Barba, C., Burkhardt, C., Boulouchos, K., and Bargende, M. A Phenomenological Model for Heat Release Rate Prediction in High-Speed DI Diesel Engine with Common Rail Injection. *SAE Paper 2000-01-2933*, 2000.

[11] Barths, H., Pitsch, H., and Peters, N. 3D Simulation of DI Diesel Combustion and Pollutant Formation Using a Two-Component Reference Fuel. *Oil Gas Sci. Techn.*, 54(2):233–244, 1999.

[12] Barths, H., Hasse, C., Bikas, G., and Peters, N. Simulation of Combustion in DI Diesel Engines Using an Eulerian Particle Flamelet Model. *Proc. Comb. Inst.*, 28:1161–1168, 2000.

[13] Battin-Leclerc, F. Detailed Chemical Kinetic Models for the Low-Temperature Combustion of Hydrocarbons with Application to Gasoline and Diesel Fuel Surrogates. *Prog. Energy Combust. Sci.*, 34(4):440–498, 2008.

[14] Baumgarten, C. *Mixture Formation in Internal Combustion Engines*. Heat, Mass Transfer. Springer-Verlag, Berlin-Heidelberg, 2006.

[15] Bellanca, R. *BlueBellMouse-A Tool for Kinetic Model Development*. PhD thesis, Lund University, 2004.

[16] Bhave, A., Balthasar, M., Kraft, M., and Mauss, F. Analysis of a Natural Gas Fuelled Homogeneous Charge Compression Ignition Engine with Exhaust Gas Recirculation using a Stochastic Reactor Model. *SAE Int. J. Engines*, 5 (1):93–103, 2004.

[17] Bhave, A., Kraft, M., Montrosi, L., and Mauss, F. Modelling a Dual-Fuelled Multi Cylinder HCCI Engine Using a PDF Based Engine Cycle Simulation. *SAE Paper 2004-01-0561*, 2004.

[18] Bhave, A., Kraft, M., Mauss, F., Oakley, A., and Zhao, H. Evaluating the EGR-AFR Operating Range of HCCI Engine. *SAE Paper 2005-01-0161*, 2005.

[19] Bird, R., Stewart, W., and Lightfoot, E. *Transport Phenomena*. John Wiley & Sons, 2006.

[20] Bordet, N., Caillol, C., Higelin, P., and Talon, V. A Physics and Tabulated Chemistry Based Compression Ignition Combustion Model: From Chemistry Limited to Mixing Limited Combustion Modes. *Oil Gas Sci. Technol.-Rev. IFP*, 2011.

[21] Bracco, F. Introducing a New Generation of More Detailed and Informative Combustion Models. *SAE Paper 71174*, 1974.

[22] Brunt, M. and Platts, K. Calculation of Heat Release in Direct Injection Diesel Engines. *SAE Paper 1999-01-0187*, 1999.

[23] Brunt, M., Rai, H., and Emtage, A. The Calculation of Heat Release Energy from Engine Cylinder Pressure Data. *SAE Paper 981052*, 1998.

[24] Cao, R., Wang, H., and Pope, S. The effect of Mixing Models in PDF Calculations of Piloted Jet Flames. *Proc. Comb. Inst.*, 31(1):1543–1550, 2007.

[25] Cengel, Y. and Boles, M. *Thermodynamics: An Engineering Approach*. Mcgraw-Hill, 2001.

[26] Chen, R. and Liu, Z. Multi-Zone Kinetic Model of Controlled Auto Ignition Combustion. *SAE Paper 2009-01-0673*, 2009.

[27] Chmela, F. and Orthaber, G. Rate of Heat Release Prediction for Direct Injection Diesel Engines Based on Purely Mixing Controlled Combustion. *SAE Paper 1999-01-0186*, 1999.

[28] Curl, R. Dispersed Phase Mixing: I. Theory and Effects in Simple Reactors. *AlChE J.*, 9(2):175–181, 1963.

[29] Davidson, P. *Turbulence: An Introduction for Scientists and Engineers.* Oxford University Press, 2004.

[30] Dec, J. Advanced Compression-ignition Engines–Understanding the In-cylinder Processes. *Proc. Combust. Inst.*, 32(2):2727–2742, 2009.

[31] Dembinski, H. *In-cylinder Flow Characterisation of Heavy Duty Diesel Engines Using Combustion Image Velocimetry.* PhD thesis, KTH Royal Institute of Technology, 2014.

[32] Diesel-Engine Management. Robert Bosch GmbH, 2005.

[33] Diesel Fuels Technical Review. `http://www.chevronwithtechron.ca/products/documents/Diesel_Fuel_Tech_Review.pdf/`, Retrieved February 16, 2013.

[34] Directive 70/220/EEC and Its Amendment and Adaptation. `http://eur-lex.europa.eu/en/index.htm/`, Retrieved January 30, 2013.

[35] Dopazo, C. Probability Density Function Approach for a Turbulent Axisymmetric Heated Jet. Centerline Evolution. *Phys. of Fluids*, 18(4):397–404, 1975.

[36] Dopazo, C. Relaxation of Initial Probability Density Functions in the Turbulent Convection of Scalar Fields. *Phys. of Fluids*, 22(1):20–30, 1979.

[37] Dopazo, C. Recent Developments in PDF Methods. In Libby, P. and Williams, F., editors, *Turbulent Reacting Flows*, pages 375–474. Academic Press Inc, 1994.

[38] Dulbecco, A., Lafossas, F., Mauviot, G., and Poinsot, T. A New 0D Diesel HCCI Combustion Model Derived from a 3D CFD Approach with Detailed Tabulated Chemistry. *Oil Gas Sci. Technol.-Rev. IFP*, 64(3):259–284, 2009.

[39] Echekki, T. and Mastorakos, E. *Turbulent Combustion Modeling: Advances, New Trends and Perspectives.* Springer Science Business Media B.V., 2011.

[40] Emission Standards. `http://www.dieselnet.com/standards/eu/ld.php/`, Retrieved January 30, 2013.

[41] Farrell, J., Cernansky, N., Dryer, F., Friend, D., Hergart, C., Law, C., McDavid, R., Mueller, C., Patel, A., and Pitsch, H. Development of an Experimental Database and Kinetic Models for Surrogate Diesel Fuels. *SAE Paper 2007-01-0201*, 2007.

[42] Fenimore, C. Studies of Fuel-Nitrogen Species in Rich Flame Gases. *Proc. Combust. Inst.*, 17(1):661–670, 1979.

[43] Fereday, D., Haynes, P., Wonhas, A., and Vassilicos, J. Scalar Variance Decay in Chaotic Advection and Batchelor-Regime Turbulence. *Phys. Rev. E*, 65: 035301, 2002.

[44] Fieweger, K., Blumenthal, R., and Adomeit, G. Self-Ignition of SI Engine Model Fuels: A Shock Tube Investigation at high Pressure. *Combust. Flame*, 109(4):599–619, 1997.

[45] Fogler, H. *Elements of Chemical Reaction Engineering*. Pearson Education Inc., 2006.

[46] Foster, D. An Overview of Zero-Dimensional Thermodynamics Models for IC Engine Data Analysis. *SAE Paper 852070*, 1985.

[47] Fox, R. Computational Methods for Turbulent Reacting Flows in the Chemical Process Industry. *Rev. I. Fr. Petrol*, 51(2):215–243, 1996.

[48] Franken, T. Simulation of Soot Formation from Diesel Engines using 0D Stochastic Reactor Model. Master's thesis, Brandenburg University of Technology, 2013.

[49] Frenklach, M. Reaction Mechanism of Soot Formation in Flames. *Phys. Chem. Chem. Phys.*, 4(11):2028–2037, 2002.

[50] Frenklach, M. Method of Moments with Interpolative Closure. *Chem. Eng. Sci.*, 57(12):2229–2239, 2002.

[51] Frenklach, M. and Wang, H. Detailed Modeling of Soot Particle Nucleation and Growth. *Proc. Comb. Inst.*, 23(1):1559–1566, 1991.

[52] Frenklach, M. and Wang, H. Soot Formation in Combustion: Mechanisms and Models. In Bockhorn, H., editor, *Detailed Mechanism and Modeling of Soot Particle Formation*, pages 165–190. Springer-Verlag, 1994.

[53] Gatowski, J., Balles, E., Chun, K., Nelson, F.E., Ekchian, J.A., and Heywood, J.B. Heat Release Analysis of Engine Pressure Data. *SAE Paper 841359*, 1984.

[54] Gogan, A., Sundén, B., Lehtiniemi, H., and Mauss, F. Stochastic Model for the Investigation of the Influence of Turbulent Mixing on Engine Knock. *SAE Paper 2004-01-2999*, 2004.

[55] Gogan, A. *Full Cycle Engine Simulations with Detailed Chemistry*. PhD thesis, Lund University, 2006.

[56] Goldberg, D. *Genetic Algorithms in Search, Optimization and Machine Learning*. Addison-Wesley Longman Publishing Co., Inc., Boston, MA, USA, 1st edition, 1989. ISBN 0201157675.

[57] Grill, M., Bargende, M., Rether, D., and Schmid, A. Quasi-Dimensional Modeling of CI-Combustion with Multiple Pilot- and Post Injection. *SAE Paper 2010-01-0150*, 2010.

[58] Hanjalić, K. *Turbulence and Transport Phenomena Modelling and Simulation.* Lecture Notes, 2006.

[59] Haworth, D. Progress in Probability Density Function Methods for Turbulent Reacting Flows. *Prog. Energy Combust. Sci.*, 36(2):168–259, 2010.

[60] Hentschel, W., Schindler, K., and Haahtela, O. European Diesel Research IDEA-Experimental Results from DI Diesel Engine Investigations. *SAE Paper 941954*, 1994.

[61] Heywood, J. *Internal Combustion Engine Fundamentals.* McGraw-Hill, New York, 1988.

[62] Hilbert, R., Tap, F., El-Rabii, H., and Thévenin, D. Impact of Detailed Chemistry and Transport Models on Turbulent Combustion Simulations. *Prog. Energy Combust. Sci.*, 30(1):61–117, 2004.

[63] Hilbig, M., Seidel, L., Wang, X., Mauss, F., and Zeuch, T. Computer Aided Detailed Mechanism Generation for Large Hydrocarbons: n-Decane. *Proceedings of the 23rd International Colloquium on the Dynamics of Explosions and Reactive Systems (ICDERS), Irvine, USA, July 24-29*, 2011.

[64] Hiroyasu, H. and Kadota, T. Models for Combustion and Formation of Nitric Oxide and Soot in Direct Injection Diesel Engines. *SAE Paper 760129*, 1976.

[65] Hiroyasu, H., Yosimatsu, A., and Arai, M. Mathematical Model for Predicting the Rate of Heat Release and Exhaust Emissions in IDI Diesel Engines. *Proc. I Mech E Conf. C102/82, p.207-2013*, 1982.

[66] James A. Miller, J. and Bowman, C. Mechanism and Modeling of Nitrogen Chemistry in Combustion. *Prog. Energy Combust. Sci.*, 15(4):287–338, 1989.

[67] Janicka, J., Kolbe, W., and Kollmann, W. Closure of the Transport-Equation for the Probability Density-Function of Turbulent Scalar Fields. *J. Non-Equilib. Thermodyn.*, 4(1):47–66, 1979.

[68] Jay, S. and Colin, O. A Variable Volume Approach of Tabulated Detailed Chemistry and its Applications to Multidimensional Engine Simulations. *Proc. Comb. Inst.*, 33(2):3065–3072, 2011.

[69] Jones, W. The Joint Scalar Probability Density Function. In Launder, B. and Sandham, N., editors, *Closure Strategies for Turbulent and Transitional Flows*, pages 582–621. Cambridge University Press, 2002.

[70] Jung, D. and Assanis, D. Multi-Zone DI Spray Combustion Model for Cycle Simulation Studies of Engine Performances and Emissions. *SAE Paper 2001-01-1246*, 2001.

[71] Kamimoto, T. and Bae, M. High Combustion Temperature for the Reduction of Particulate in Diesel Engines. *SAE Paper 880423*, 1988.

[72] Kennedy, I. Models of Soot Formation and Oxidation. *Prog. Energy Combust. Sci.*, 23(2):95–132, 1997.

[73] Kim, S., Wakisaka, T., and Aoyagi, Y. A Numerical Study of the Effects of Boost Pressure and Exhaust Gas Recirculation Ratio on the Combustion Process and Exhaust Emissions in a Diesel Engine. *Int. J. Engine Res.*, 8(2): 147–162, 2007.

[74] Kitamura, T., Ito, T., Senda, J., and Fujimoto, H. Mechanism of Smokeless Diesel Combustion with Oxygenated Fuels Based on the Dependence of the Equivalence Ratio and Temperature on Soot Particle Formation. *Int. J. Engine Res.*, 3(4):223–248, 2002.

[75] Kook, S. and Pickett, L. Soot Volume Fraction and Morphology of Conventional and Surrogate Jet Fuel Sprays at 1000-K and 6.7-MPa Ambient Conditions. *Proc. Comb. Inst.*, 33(2):2911–2918, 2011.

[76] Kraft, M. *Stochastic Modelling of Turbulent Reacting Flow in Chemical Engineering.* VDI Verlag, 1998.

[77] Kraft, M., Maigaard, P., Mauss, F., Christensen, M., and Johansson, B. Investigation of Combustion Emissions in a Homogeneous Charge Compression Injection Engine: Measurements and a new Computational Model. *Proc. Comb. Inst.*, 28(1):1195–1201, 2000.

[78] Krieger, R. and Borman, G. The Computation of Apparent Heat Release in Internal Combustion Engines. *Proceedings of the Winter Annual Meeting and Energy Systems Exposition (ASME 66-WA/DGP-4), New York, USA, November 27-December 1*, 1966.

[79] Kuo, K. *Principles of Combustion.* John Wiley & Sons, 1986.

[80] Kuo, K. and Acharya, R. *Fundamentals of Turbulent and Multiphase Combustion.* John Wiley & Sons, Inc., 2012.

[81] Lafossas, F., Marbaix, M., and Mengazzi, P. Development and Application of a 0D D.I. Diesel Combustion Model for Emission Prediction. *SAE Paper 2007-01-1841*, 2007.

[82] Lavoie, G., Heywood, J., and Keck, J. Experimental and Theoretical Study of Nitric Oxide Formation in Internal Combustion Engines. *Combust. Sci. Techn.*, 1(4):313–326, 1970.

[83] Lehtiniemi, H., Mauss, F., Balthasar, M., and Magnusson, I. Modeling Diesel Spray Ignition Using Detailed Chemistry with a Progress Variable Approach. *Combust. Sci. Techn.*, 178(10-11):1977–1997, 2006.

[84] Lehtiniemi, H., Zhang, Y., Rawat, R., and Mauss, F. Efficient 3-D CFD Combustion Modeling with Transient Flamelet Models. *SAE Paper 2008-01-0957*, 2008.

[85] Libby, P. and Williams, F. *Turbulent Reacting Flows.* Academic Press, 1994.

[86] Liu, S., Li, H., Gatts, T. Liew, C., Wayne, S., Thompson, G., Clark, N., and Nuszkowski, J. An investigation of NO2 Emissions from a Heavy-duty Diesel Engine Fumigated with H2 and Natural gas. *Combust. Sci. Techn.*, 184(12): 2008–2035, 2012.

[87] Loge AB. *LOGEsoft v1.04, LOGEengine v1.04; User Manuals*, 2013.

[88] Løvås, T. *Automatic Reduction Procedures for Chemical Mechanisms in Reactive Systems*. PhD thesis, Lund University, 2002.

[89] Lu, X., Han, D., and Huang, Z. Fuel Design and Management for the Control of Advanced Sompression-Ignition Combustion Modes. *Prog. Energy Combust. Sci.*, 37(3):741–783, 2011.

[90] Maigaard, P., Mauss, F., and Kraft, M. Homogenous Charge Compression Ignition Engine: A Simulation Study on the Effect of Inhomogeneities. *J. Eng. Gas Turb. Power*, 125(2):466–471, 2003.

[91] Mastorakos, E. and Cant, R. *An Introduction to Turbulent Reacting Flows*. Imperial College Press, 2008.

[92] Mauß, F. *Entwicklung eines kinetischen Modells der Rußbildung mit schneller Polymerisation*. PhD thesis, RWTH Aachen, 1998.

[93] Mauss, F., Keller, D., and Peters, N. A Lagrangian Simulation of Flamelet Extinction and Re-ignition in Turbulent Jet Diffusion Flames. *Proc. Combust. Inst.*, 23(1):693–698, 1991.

[94] Mauss, F., Schäfer, T., and Bockhorn, H. Inception and Growth of Soot Particles in Dependence on the Surrounding Gas Phase. *Combust. Flame*, 99 (3-4):697–705, 1994.

[95] Merker, G., Schwarz, C., Stiesch, G., and Otto, F. *Simulating Combustion. Simulation of Combustion and Pollutant Formation for Engine-Development*. Springer-Verlag Berlin Heidelberg, 2006.

[96] Merker, G., Schwarz, C., and Teichmann, R. *Combustion Engines Development: Mixture Formation, Combustion, Emissions and Simulation*. Springer-Verlag Berlin Heidelberg, 2012.

[97] Meyer, D. and Jenny, P. Micromixing Models for Turbulent Flows. *J. Comput. Phys.*, 228:1275–1293, 2009.

[98] Mitchell, P. and Frenklach, M. Particle Aggregation with Simultaneous Surface Growth. *Phys. Rev. E*, 67(6):Art. No. 061407 Part 1, 2003.

[99] Müller, E. and Zillmer, M. Modeling of Nitric Oxide and Soot Formation in Diesel Engine Combustion. *SAE Paper 982457*, 1998.

[100] Mollenhauer, K. and Tschöke, H. *Handbuch Dieselmotoren*. VDI-Buch. Springer-Verlag Berlin Heidelberg, 2007.

[101] Montorsi, L., Mauss, F., Bianchi, G., Bhave, A., and Kraft, M. Analysis of the HCCI Combustion of a Turbocharged Truck Engine Using a Stochastic Reactor Model. *Proc. ASME-ICE Conf.*, 2002(46628):97–108, 2002.

[102] Mosbach, S., Su, H., Kraft, M., Bhave, A., Mauss, F., Wang, Z., and Wang, J.X. Dual Injection HCCI Engine Simulation using a Stochastic Reactor Model. *Int. J. Engine Res.*, 8(1):41–50, 2007.

[103] Murphy, M., Taylor, J., and McCormick, R. Compendium of experimental cetane number data. National Renewable Energy Laboratory,http://www.nrel.gov/docs/legosti/old/36805.pdf/, Retrieved January 15, 2014.

[104] Nakov, G., Mauss, F., Wenzel, P., and Krüger, C. Application of a Stationary Flamelet Library Based CFD Soot Model for Low-NOx Diesel Combustion. *Proceedings of THIESEL2010 Conference on Thermo- and Fluid Dynamic Processes in Diesel Engines, Valencia, Spain, September 14-17*, 2010.

[105] Nakov, G., Mauss, F., Wenzel, P., Steiner, R., Krüger, C., Zhang, Y., Rawat, R., Borg, A., Perlman, C., Fröjd, K., and Lehtiniemi, H. Soot Simulation under Diesel Engine Conditions Using a Flamelet Approach. *SAE Int. J. Engines*, 2(2):89–104, 2010.

[106] Nooren, P., Wouters, H., Peeters, T., Roekaerts, D., Maas, U., and Schmidt, D. Monte Carlo PDF Modelling of a Turbulent Natural-Gas Diffusion Flame. *Combust. Theory Model.*, 1(1):79–96, 1997.

[107] Owen, K. and Coley, C. *Automotive Fuels Reference Book*. Society of Automotive Engineers, Inc., 1995.

[108] Papula, L. *Mathematische Formelsammlung foer Ingenieure und Naturwissenschaftler*. Vieweg+Teubner, 2009.

[109] Pasternak, M. and Mauss, F. Simulation of Fuel Effects under Diesel-Engine Conditions using 0D-Fuel-Test Bench. *Berichte zur Energie- und Verfahrenstechnik (BEV) 13.1, 337-346, ISBN 3-931901-87-4*, 2013.

[110] Pasternak, M. and Mauss, F. Aspects of Diesel Engine In-Cylinder Processes Simulation Using 0D Stochastic Reactor Model. *Proceedings of the 1st International Conference on Engine Processes (ISBN 78-3-8169-3222-2), Berlin, Germany, June 6-7*, 2013.

[111] Pasternak, M., Mauss, F., and Bensler, H. Diesel Engine Cycle Simulations with a Reduced Set of Modeling Parameters Based on Detailed Kinetic. *SAE Paper 2009-01-0676*, 2009.

[112] Pasternak, M., Mauss, F., and Lange, F. Time Dependent Based Mixing Time Modelling for Diesel Engine Combustion Simulations. *Proceedings of the 23rd International Colloquium on the Dynamics of Explosions and Reactive Systems (ICDERS), Irvine, USA, July 24-29*, 2011.

[113] Pasternak, M., Mauss, F., Janiga, G., and Thévenin, D. Self-Calibrating Model for Diesel Engine Simulations. *SAE Paper 2012-01-1072*, 2012.

[114] Pasternak, M., Mauss, F., Matrisciano, A., and Seidel, L. Simulation of Diesel Surrogate Fuels Performance under Engine Conditions using 0D Engine – Fuel Test Bench. *Proceedings of the 8th International Conference on Modelling and Diagnostics for Advanced Engine Systems (COMODIA), Fukuoka, Japan, July 23-26*, 2012.

[115] Pasternak, M., Mauss, F., Perlman, C., and Lehtiniemi, H. Aspects of 0D and 3D Modeling of Soot Formation for Diesel Engines. *Combust. Sci. and Tech.*, pages 1517–1535, 2014.

[116] Perlman, C., Fröjd, K., Seidel, L., Tunér, M., and Mauss, F. A Fast Tool for Predictive IC Engine In-Cylinder Modelling with Detailed Chemistry. *SAE Paper 2012-01-1074*, 2012.

[117] Peters, N. *Fifteen Lectures on Laminar and Turbulent Combustion.* ERCOFTAC Summer School, Aachen, 1992.

[118] Peters, N. *Turbulent Combustion.* Cambridge University Press, 2000.

[119] Pirker, G., Chmela, F., and Wimmer, A. ROHR Simulation for DI Diesel Engines Based on Sequential Combustion Mechanisms. *SAE Paper 2006-01-0654*, 2006.

[120] Pitsch, H., Wan, Y., and Peters, N. Numerical Investigation of Soot Formation and Oxidation under Diesel Conditions. *SAE Paper 952357*, 1995.

[121] Pitsch, H., Barths, H., and Peters, N. Three-Dimensional Modeling of NOx and Soot Formation in DI Diesel Engines Using Detailed Chemistry Based on the Interactive Flamelet Approach. *SAE Paper 962057*, 1996.

[122] Pitsch, H., Riesmeier, E., and Peters, N. Unsteady Flamelet Modeling of Soot Formation in Turbulent Diffusion Flames. *Combust. Sci. Techn.*, 158(1): 389–406, 2000.

[123] Pitz, W. and Mueller, C. Recent Progress in the Development of Diesel Surrogate Fuels. *Prog. Energy Combust. Sci.*, 37(3):330–350, 2011.

[124] Poinsot, T. and Veynante, D. *Theoretical and Numerical Combustion.* Edwards, Philadelphia, PA, 2005.

[125] Pope, S. An Improvend Turbulent Mixing Model. *Combust. Sci. Techn.*, 28: 131–145, 1982.

[126] Pope, S. Pdf Methods for Turbulent Reactive Flows. *Prog. Energy Combust. Sci.*, 11(2):119–192, 1985.

[127] Pope, S. The PDF Method for Turbulent Combustion. *CFD Symposium on Aeropropulsion, NASA Lewis*, 1991.

[128] Prabhu, S., Li, H., Miller, S., and Cernansky, N. The Effect of Nitric Oxide on Autoignition of a Primary Reference Fuel Blend in a Motored Engine. *SAE Paper 932757*, 1991.

[129] Ratner, B. *Statistical and Machine-Learning Data Mining: Techniques for Better Predictive Modeling and Analysis of Big Data, Second Edition.* Taylor & Francis, 2011.

[130] Ren, Z. and Pope, S. An Investigation of the Performance of Turbulent Mixing Models. *Combust. Flame*, 136:208–216, 2004.

[131] Rezaei, R., Eckert, P., Seebode, J., and Behnk, K. Zero-Diemsional Modelling of Combustion and Heat Release Rate in DI Diesel Engines. *SAE Int. J. Engines*, 5(3), 2012.

[132] Roekaerts, D. Reacting Flows and Probability Density Function Methods. In Launder, B. and Sandham, N., editors, *Closure Strategies for Turbulent and Transitional Flows*, pages 328–337. Cambridge University Press, 2002.

[133] Rychter, T. and Teodorczyk, A. *Modelowanie matematczne roboczego cyklu silnika tłokowego*. PWN Warszawa, 1990.

[134] Samuelsson, K. Development and Validation of a Fuel Injection Model for the SRM Code. Master's thesis, Lund University, 2004.

[135] Samuelsson, K., Gogan, A., Netzell, K., Lehtiniemi, H., Sunden, B., and Mauss, F. Modeling Diesel Engine Combustion and Pollutant Formation Using a Stochastic Reactor Model Approach. *Proceedings of the International Symposium "Towards Clean Diesel Engines", Lund, Sweden, June 2-3*, 2005.

[136] Schindler, K. Integrated Diesel European Action (IDEA): Study of Diesel Combustion. *SAE Paper 920591*, 1992.

[137] Schuetz, C. and Frenklach, M. Nucleation of Soot: Molecular Dynamics Simulations of Pyrene Dimerization. *Proc. Combust. Inst.*, 29:2307–2314, 2002.

[138] Seidel, L., Bruhn, F., Ahmed, S., Moreác, G., Zeuch, T., and Mauss, F. The Comprehensive Modelling of n-Heptane / iso-Octane Oxidation by a Skeletal Mechanism using the Chemistry Guided Reduction Approach. *Proceedings of the 22nd International Colloquium on the Dynamics of Explosions and Reactive Systems (ICDERS), Minsk, Belarus,July 27-31*, 2009.

[139] Shenghua, L., Hwang, J., Park, J., Kim, M., and Chae, J. Multizone Model for DI Diesel Engine Combustion and Emissions. *SAE Paper 1999-01-2926*, 1999.

[140] Smallbone, A., Bhave, A., Coble, A., Mosbach, S., Kraft, M., and McDavid, R. Identifying Optimal Operating Points in Terms of Engineering Constraints and Regulated Emissions in Modern Diesel Engines. *SAE Paper 2011-01-1388*, 2011.

[141] Smith, G., Golden, D., Frenklach, M., Eiteener, B., Goldenberg, M., Bowman, C., Hanson, R., Gardiner, W., Lissianski, V., and Qin, Z. GRI-Mech 3.0. http://www.me.berkeley.edu/gri_mech/, 2000.

[142] Smith, J., Simmie, J., and Curran, H. Detailed Chemical Kinetic Models for the Combustion of Hydrocarbon Fuels. *Prog. Energy Combust. Sci.*, 29(6): 599–634, 2003.

[143] Srinivas, M. and Patnaik, L. Genetic algorithms: A survey. *IEEE Computer*, 27(6):17–26, 1994.

[144] Steiner, R., Bauer, C., Krüger, C., Otto, F., and Maas, U. 3-D Simulation of DI-Diesel Combustion Applying a Progress Variable Approach Accounting for Complex Chemistry. *SAE Paper 2004-01-0106*, 2004.

[145] Stenlåås, O., Gogan, A., Egnell, R., Sundén, B., and Mauss, F. The Influence of Nitric Oxide on the Occurrence of Autoignition in the End Gas of Spark Ignition Engines. *SAE Paper 2002-01-2699*, 2002.

[146] Stiesch, G. *Modelling Engine Spray and Combustion Processes.* Springer-Verlag Berlin Heidelberg, 2003.

[147] Stiesch, G. and Merker, G. A Phenomenological Heat Release Model for Direct Injection Diesel Engines. *Proceedings of the 22nd CIMAC Congress, 423-429,Copenhagen,* 1998.

[148] Stiesch, G. and Merker, G. A Phenomenological Model for Accurate and Time Efficient Prediction the Rate of Heat Release and Exhaust Emissions in Direct-Injection Diesel Engine. *SAE Paper 1999-01-1535,* 1999.

[149] Su, H., Mosbach, S., Kraft, M., Bhave, A., Kook, S., and Bae, C. Two-stage Fuel Direct Injection in a Diesel Fuelled HCCI Engine. *SAE Paper 2007-01-1880,* 2007.

[150] Su, H., Vikhansky, A., Mosbach, S., Kraft, M., Bhave, A., Kim, K., Kobayashi, T., and Mauss, F. A Computational Study of an HCCI Engine with Direct Injection During Gas Exchange. *Combust. Flame,* 147(1-2):118–132, 2006.

[151] Subramaniam, S. and Pope, S. A Mixing Model for Turbulent Reactive Flows Based on Euclidean Minimum Spanning Trees. *Combust. Flame,* 115 (4):487Ű514, 1998.

[152] Tao, F., Reitz, R., and Foster, D. Revisit of Diesel Reference Fuel (n-Heptane) Mechanism Applied to Multidimensional Diesel Ignition and Combustion Simulations. *Proceedings of the International Multidimensional Engine Modeling User's Group Meeting, Detroit, USA, April 15,* 2007.

[153] Tennekes, H. and Lumley, J. *A First Course in Turbulence.* MIT Press, 1972.

[154] The Kyoto Protocol. `http://unfccc/int/kyoto_protocol/items/2830.php/`, Retrieved January 30, 2013.

[155] Tree, D. and Svensson, K. Soot Processes in Compression Ignition Engines. *Prog. Energy Combust. Sci.,* 33(3):272–309, 2007.

[156] Tsurushima, T. A New Skeletal PRF Kinetic Model for HCCI Combustion. *Proc. Comb. Inst.,* 32(2):2835–2841, 2009.

[157] Tunér, M., Pasternak, M., Mauss, F., and Bensler, H. A PDF-Based Model for Full Cycle Simulations of Direct Injected Engines. *SAE Paper 2008-01-1606,* 2008.

[158] Turns, S. *An Introduction to Combustion. Concepts and Applications.* McGraw-Hill, International Edition, 2006.

[159] United States Environmental Protection Agency. Air Pollutants. `http://www.epa.gov/airquality/urbanair/`, Retrieved January 30, 2013.

[160] Valino, L. and Dopazo, C. A Binomial Langevin Model for Turbulent Mixing. *Phys. Fluids A: Fluid Dynamics,* 3(12):3034–3037, 1991.

[161] Veynante, D. and Vervisch, L. Turbulent Combustion Modeling. *Prog. Energy Combust. Sci.,* 28(3):193–266, 2002.

[162] Wallington, T., Kaiser, E., and Farrell, J. Fuels and Internal Combustion Engines: A Chemical Perspective. *Chem. Soc. Rev.*, 35(4):335–347, 2006.

[163] Warnatz, J., Maas, U., and Dibble, R. *Combustion. Physical and Chemical Fundamentals, Modeling and Simulations, Experiments, Pollutant Formation.* Springer-Verlag Berlin Heidelberg, 2001.

[164] Weisser, G. *Modelling of Combustion and Nitric Oxide Formation for Medium-Speed DI Diesel Engines: A Comparative Evaluation of Zero- and Three–Dimensional Approaches.* PhD thesis, Swiss Federal Institute of Technology Zurich, 2001.

[165] Wenzel, S. *Modellierung der Russ– und NOx– Emissionen des Dieselmotors.* PhD thesis, TU Magdeburg, 2006.

[166] Wilcox, D. *Turbulence Modeling for CFD.* DCW Industries, 1993.

[167] Wolfram MathWorld. Generalized Functions. `http://mathworld.wolfram.com/topics/GeneralizedFunctions.html/`, Retrieved July 1, 2013.

[168] Wolfrum, J. Bildung von Stickstoffoxiden bei der Verbrennung. *Chemie-Ingenieur-Technik.*, 44:656–659, 1972.

[169] Woschni, G. A Universally Applicable Equation for the Instantaneous Heat Transfer Coefficient in the Internal Combustion Engine. *SAE Paper 670931*, 1967.

[170] Xu, J. and Pope, S. PDF Calculations of Turbulent Nonpremixed Flames with Local Extinction. *Combust. Flame*, 123(3):281–307, 2000.

[171] You, X., Egolfopoulos, F., and Wang, H. Detailed and Simplified Kinetic Models of n-Dodecane Oxidation: The Role of Fuel Cracking in Aliphatic Hydrocarbon Combustion. *Proc. Comb. Inst.*, 32(1):403–410, 2009.

[172] Zeldovich, Y. The Oxidation of Nitrogen in Combustion Explosions. *Acta Physicochimica U.R.S.S.*, 21:577–628, 1946.

[173] Zeuch, T., Moreác, G., Ahmed, S., and Mauss, F. A Comprehensive Skeletal Mechanism for the Oxidation of n-Heptane Generated by Chemistry-Guided Reduction. *Combust. Flame*, 155(4):651–674, 2008.

Appendices

The appendices contain technical highlights of Diesel engines and operating points that were used as calculation test cases for modelling and simulations carried out in this work.

A Test Engines

Four Diesel engines were used as test calculations. They are respectively denoted as **Engine A, Engine B, Engine C and Engine D**.

Table A.1: Technical specification of test engines

Parameter	Engine A	Engine B	Engine C	Engine D
Bore [mm]	83.0	81.0	81.0	81.0
Stroke [mm]	99.5	95.5	95.5	95.5
Compression ratio [–]	16.2	16.3	18.0	15.5
Fuel type [–]	Diesel	n-heptane	Diesel	Diesel

All engines are Diesel with direct fuel injection. The engines A, B and D are single cylinder research engines and engine C is a four cylinder serial engine.

The engine technical data (Table A.1) and corresponding operating conditions (Table B.1) have been extracted from the previously presented works: [104] (Engine A) [112, 113] (Engine B) and [110] (Engine C).

B Engine Operating Points

The appendix describes engine operating points investigated. Emphasis is placed on six points described in Table B.1 for which more detailed results are presented in this work. The selected operating points correspond to engines from Table A.1 and are referred to as *Cases*. They differ by speed, load, EGR rate and fuel injection timing. For all engine cases discussed, the measured exhaust emissions used as reference for models validation are given in mass-based concentration, unless stated otherwise.

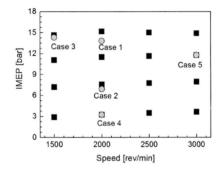

Figure B.1: Reference engine operating points. Squares: operating points used for the study on mixing time parametrisation. Circles: representative operating points that are discussed in more detail in this report

Table B.1: Specification of engine operating points

Parameter	Case 1 Engine A	Case 2 Engine B	Case 3 Engine C	Case 4 Engine D	Case 5 Engine D
Speed [1/min]	2000	2000	1500	2000	3000
IMEP [bar]	13.8	6.90	14.3	3.24	11.7
Mass of fuel injected [mg]	44.0	17.2	42.0	10.0	30.0
Start of injection [°CA ATDC]	-3.0	-2.0	-10.0	0.0	-7.0
Fuel injection mode [–]	single	single	double	single	single
Equivalence ratio [–]	0.75	0.55	0.69	0.18	0.60
EGR [–]	0.27	0.33	0.0	0.20	0.20